Advances in Intelligent Systems and Computing

Volume 728

Series editor

Janusz Kacprzyk, Polish Academy of Sciences, Warsaw, Poland
e-mail: kacprzyk@ibspan.waw.pl

The series "Advances in Intelligent Systems and Computing" contains publications on theory, applications, and design methods of Intelligent Systems and Intelligent Computing. Virtually all disciplines such as engineering, natural sciences, computer and information science, ICT, economics, business, e-commerce, environment, healthcare, life science are covered. The list of topics spans all the areas of modern intelligent systems and computing.

The publications within "Advances in Intelligent Systems and Computing" are primarily textbooks and proceedings of important conferences, symposia and congresses. They cover significant recent developments in the field, both of a foundational and applicable character. An important characteristic feature of the series is the short publication time and world-wide distribution. This permits a rapid and broad dissemination of research results.

More information about this series at http://www.springer.com/series/11156

Fabio Leuzzi · Stefano Ferilli
Editors

Traffic Mining Applied to Police Activities

Proceedings of the 1st Italian Conference for the Traffic Police (TRAP-2017)

 Springer

Editors
Fabio Leuzzi
Italian National Police
Rome
Italy

Stefano Ferilli ⓘ
University of Bari
Bari
Italy

ISSN 2194-5357   ISSN 2194-5365   (electronic)
Advances in Intelligent Systems and Computing
ISBN 978-3-319-75607-3   ISBN 978-3-319-75608-0   (eBook)
https://doi.org/10.1007/978-3-319-75608-0

Library of Congress Control Number: 2018932179

Printed on acid-free paper

This Springer imprint is published by the registered company Springer International Publishing AG
part of Springer Nature
The registered company address is: Gewerbestrasse 11, 6330 Cham, Switzerland

Foreword

Applied research. These are perfect keywords to describe our times, efforts, needs, and prospects of human development. For the first time, with TRAP-2017, the Italian Traffic Police—I would say the Italian National Police in general—has organized a scientific conference in which the competences, specialties, and topics of a police force constitute the observation ground of applied research.

Such an initiative might sound unusual for a Public Administration, whereas one might think that the organizers are about a decade late. Whatever the reader thinks about this, the main message we want to convey concerns our new look.

The Italian National Police continuously strives to fully understand the various phenomena, to fight crime and ensure safety and security. Nowadays, however, prevention and repression are not enough to appropriately tackle the new challenges. The great amount of data around us cannot be ignored to successfully and timely meet our objectives. Therefore, our aim is to start a new age in our history, an age centered on technical and sophisticated approaches to extract information from the vast amount of data which otherwise could not be used, an age focusing on the improvement of our abilities to predict significant events. All tasks that can be no longer performed through paper and pencil only.

The safeguard of the freedom of movement, which represents the glorious past and the proud present of the Italian Traffic Police, will still and always be the core of its identity. This identity is essential to be at the forefront in facing the new challenges posed by the so-called smart cities, smart roads, by the circulation of intelligent cars, by Intelligent Transport Systems and other innovations that represent our present and future.

Let us make some brief considerations on the main difference between high school and university. The former aims at providing students with valuable bases upon which specialized studies can be carried out, basing teaching on books containing consolidated notions. University, on the other hand, is the place where the state-of-the-art and latest innovations are developed and, thanks to the research activities of professors and researchers, transferred to students.

The same can be said for the Italian National Police. This volume officially marks the moment in which the Italian Traffic Police improves its approach to knowledge and officially enters the world of research, dealing directly with the relevant academic experts in order to apply state of the art of applied research and perform its institutional activities through up-to-date knowledge, exploiting the most advanced techniques and innovative tools that research in science and engineering can provide.

I am pleased to witness and strongly support these new and strategic directions, and I am sure that these efforts will soon result in tangible outcomes.

October 2017   Franco Gabrielli

Preface

The First Italian Conference on Traffic Mining applied to Police Activities (TRAP-2017) was held in Rome during October 25–26, 2017, in the context of the celebrations for the 70th Anniversary of the Italian Traffic Police (*Polizia Stradale Italiana*). Its aim was to gather data mining researchers, traffic researchers, and decision makers and provide them a common forum for discussing the development and exploitation of automatic traffic analysis systems that can detect, track and, more in general, understand the behavior of road users in order to identify criminal behaviors.

Indeed, with the increasing amount of traffic information collected through automatic number plate reading systems (NPRS), which are widely spread on Italian highways, it is highly desirable for police activities and investigations to be able to extract meaningful traffic patterns from the accumulated massive historical dataset, in order to identify potential criminal behaviors. However, analyzing traffic data for this purpose is challenging due to the huge size of the dataset and the complexity and dynamics of traffic phenomena.

Topics of interest included detection and tracking of road users and vehicles, behavior understanding of road users, automatic understanding of the environment in traffic scenarios, applications related to traffic surveillance, and vehicle accident analysis. Examples of techniques of interest are outliers detection and understanding, clustering and conceptual clustering, process mining, inductive logic programming, deep learning and classification.

To allow interested researchers to work on real-world data that are hardly available to the wide public, the organizers provided a dataset reporting transit data recorded using several gates spread along a limited area of Italy, in which gates are homogeneously distributed. As an option, authors had the possibility to propose contributions specifically focused on this dataset.

With the aim of unifying the way experimental results are evaluated and to push research forward on the development of real working systems supporting police activity, a shared task was proposed, whose main practical goal was to identify itineraries that might indicate a criminal intent. The scientists were free to define the concept of itinerary, formalizing it functionally to their proposed approach. They were free to integrate open data into the itinerary features or not. Criminal intents could be described, for instance, as follows:

– The sequential visit of service areas facing each other;
– The sequential visit of service areas in the same direction;
– Transits that are inconsistent under the space-time point of view, that might be due to cloned plates;
– Combinations of the previous cases involving several plates and possibly the same criminal organization;

and so on.

Two invited talks and nine original research papers related to the conference main topics, or to other relevant topics of interest to traffic understanding, were presented during the conference, allowing researchers and practitioners to pose problems and propose solutions, to identify common tasks of interest and to plan possible cooperation.

We would like to thank the Italian National Police for its invaluable support to the organization of this conference. Our special thanks go to the Director of the Central Directorate for the Specialties of the Italian National Police, Roberto Sgalla, and to the Director of the Italian Traffic Police, Giuseppe Bisogno. They quickly understood the importance of applied research for police activities and strongly believed that the TRAP-2017 initiative could be appreciated by researchers around the world. Grateful thanks are also due to all the sponsors for making this event possible, and to all the people who contributed to the organization and success of the conference. Let's hope this was just the beginning.

TRAP-2017

Rome, Italy   Fabio Leuzzi
Bari, Italy   Stefano Ferilli
October 2017

Organization

TRAP-2017 is organized by the Italian National Police, Traffic Police Department, Ministry of Interior.

Executive Committee

Program Chair

Stefano Ferilli, University of Bari, Italy
Ambra Gentile, Italian National Police, Italy
Salvatore Iengo, Italian National Police, Italy
Fulvio Rotella, Italian National Police, Italy

Program and Executive Chair

Fabio Leuzzi, Italian National Police, Italy

Program Committee

Dino Pedreschi, University of Pisa, Italy
Fosca Giannotti, Information Science and Technology Institute of the National Research Council at Pisa, Italy
Stefano Ferilli, University of Bari, Italy
Federico Chesani, University of Bologna, Italy
Claudia Diamantini, University of Ancona, Italy
Fabiana Lanotte, Digital Transformation Team, Italy
Giorgia Lodi, Agenzia per l'Italia Digitale, Italy
Giacinto Occhiogrosso, Italian National Police, Italy
Tommaso Fornaciari, Italian National Police, Italy

Fabio Leuzzi, Italian National Police, Italy
Salvatore Iengo, Italian National Police, Italy
Onofrio Febbraro, Italian National Police, Italy
Fulvio Rotella, Italian National Police, Italy
Emiliano Del Signore, Italian National Police, Italy

Organizing Committee

Fabio Leuzzi, Italian National Police, Italy
Onofrio Febbraro, Italian National Police, Italy

Sponsoring Institutions

Main Sponsor

Oracle, Italy

Silver Sponsor

Divitech, Italy

Partner

Autovie Venete, Italy

Contents

Part I
Invited Talks

Data and Analytics Framework. How Public Sector Can Profit from Its Immense Asset, Data

Raffaele Lillo

Abstract Public sector is rich of data, but this alone is not enough to fully exploit insights and information hidden in it. Data needs to be coupled with a team of data scientists and engineers, a big data platform and a legislative framework to make the famous "data driven decision making" actually possible. This is why the Digital Transformation Team introduced the Data and Analytics Framework.

Keywords Public administration · Data and analytics framework · Big data

1 Introduction

"Submersion" is a serious phenomenon that affects the whole of Italy. We are not, in this case, referring to the all-too-common practice of tax evasion. There is another, equally large source of capital to bring to the surface, and its mechanism of "recovery" is technological, organizational and legislative. There is a great resource that is not able to emerge and make itself useful to citizens, businesses and all public administrations because it is fragmented, scattered across different places, imprisoned by rules and practices that impede movement, sharing and optimal use. Public information is the enormous data-set that describes the realities of citizens and businesses — where and how we live, what we do—and, like the first public investments in telecommunication, represents a strategic asset to take advantage of, even with state intervention. It pertains to all of us and is necessary to start up businesses, conduct activities and access public services. It contains data the Public Administration (PA) needs when it offers services to citizens; data that can help the State identify problems more efficiently and develop better solutions; data through which citizens might get to know the actions of the State and assess its results. To tackle this challenge, The Italian Government, via its Digital Transformation Team led by Extraordinary Commissioner Diego Piacentini, is designing and developing the Data and Analytics

R. Lillo (✉)
Digital Trasformation Team, Presidenza del Consiglio dei Ministri, Italian Government, Rome, Italy
e-mail: raffaele@teamdigitale.governo.it

© Springer International Publishing AG, part of Springer Nature 2018
F. Leuzzi and S. Ferilli (eds.), *Traffic Mining Applied to Police Activities*, Advances in Intelligent Systems and Computing 728,
https://doi.org/10.1007/978-3-319-75608-0_1

Framework (DAF). The main goal of the DAF is to establish a Chief Data Office of the PA that will set the data strategy for the country and will provide once and for all infrastructure and expertise to all PA. This will translate into: an optimized data exchange between PAs, exiting from the current "siloed" and compartmentalized situation; coherent data strategy across PAs and sectors of society, and adoption of common standards; better interoperability; improved efficiency due to the provisioning of infrastructure and standard functionalities managed in a Platform as a Service fashion, so to free PA's resources to be focused on domain specific and higher value added activities.

2   Big Data Analytics for the Public Administration

Big Data analytics techniques can be used to gather and to use precious information that are usually impracticable (when not impossible) to obtain via traditional methods. The main advantages of this approach are threefold. In the first place, it allows for the usage of multiple and heterogeneous integrated data sources describing a phenomenon (the Variety of the famous Big Data Three Vs), offering multiple angles of analysis. Secondly, it introduces a different approach to traditional economic and statistical analysis, as it allows to ask the data to produce insights, instead of validating hypotheses based on a priori theoretical models. This is particularly useful for complex and interconnected phenomena for which there are "a lot" of data available (more than a human can reasonably handle; here "a lot" refers to both Volume and Variety of the Three Vs). The complexity and availability of huge amount of information is a binomial. That is quite common nowadays thanks to new technologies that allows to collect and manage data cheaply, and it is easily applicable to social activities in general. Finally, technological progress gives the possibility to collect data at a much faster pace and in a timely manner, allowing analysis and responses to be performed right after an event has happened (the Velocity of the three Vs).

A well thought data strategy coupled with established Big Data technologies proved to bring value and provide competitive advantage to corporations. Quoting Kiron, Ferguson and Prentice (2013): *"In the future, analytics will transition from simple information gatherers and maintainers to influential thought-leaders that are integrated parts of teams across organization"*. Furthermore, several studies showed a positive and significant impact on P&L of companies that embraced Big Data activities. In particular, Mckinsey (2011) showed that companies using Big Data techniques have outperformed their respective markets and have created competitive advantage, by comparing both revenues and EBITDA against their peers. Public sector is no different. Resources and capabilities needed to support the decision making processes at all levels are crucial to improve the effectiveness (and efficiency) with which PAs function and serve citizens. Furthermore, the typical scale of a government coupled with the high degree of fragmentation among PA are two aspects that marry perfectly with the horizontal scaling of Big Data architectures and the institutionalization of a central office that manages the platform and provide a uniform and coherent data strategy for all PAs.

3   Big Data and Public Policy

In analyzing the areas where Big Data analytics techniques may help Government and PAs, it is useful to follow the systematics of administrative functions as defined by Maciejewski (2017):

- **Public regulation**: putting in place policy to shape social behavior and relations by means of permits, prohibitions and orders
- **Public supervision**: monitoring activities, anomalies and irregularities detection, implementing responsive actions
- **Public service delivery**: providing services to citizens and enterprises

Public regulation comprises all the phases of the policy making process, from the identification of the need of a new policy, analysis and impact assessment, policy implementation, fine-tuning. All these phases can highly benefit from big data techniques: continuous and automated mining of "all" available data can help identify otherwise hidden patterns and shape the need for new policies or fine tune existing ones; impact assessment can be expanded to increase the number of connections with aspects of society impacted by the new policy; likewise, the monitoring and fine-tuning phases can benefit from the timeliness and richness of available information.

Public supervision is probably the area where the advantages of Big Data are most evident. In fact, analysis can make use of the widest possible sets of information available to prepare models to recognize irregularities. One of the most direct application of big data to public supervision is fraud detection applied in the fiscal context.

Finally, big data can help improve public services offered to citizens and enterprises in two ways: it supports service design by providing useful information about end-users that are fundamental to better understand needs and usage patterns, and therefore create more effective and customized services; it reuses information already collected by any public administration, minimizing the level of effort required by end-users.

4   Data and Analytics Framework for the Public Administration

The DAF is a component of the 3-year plan for the PA, published in May 2017 and signed by Italian Prime Minister, with the aim of providing the PAs with skills, infrastructure and proper regulation to allow them to make the most out of all their data. The DAF is a framework made of three macro components:

- a Big Data Platform to store in a unique repository the data of the PAs, implementing ingestion procedures to promote standardization and therefore interoperability among them. It exposes functionalities common to the Hadoop ecosystem, plus a set of (micro) services designed around use cases typical for public administrations;
- a Team of Data Experts (data scientists and data engineers) that knows how to manage and evolve the Big Data Platform, and provide support to PA on their analytics and data management activities in a consultancy fashion;
- a Regulatory Framework that institutionalizes this activity at government level, and gives to the organization that will manage the DAF the proper mandate in compliance with privacy policy.

A framework with these characteristics may be effective to successfully cope with well-known problems when it comes to using and sharing data. A typical issue arises from situations where there are multiple organizations managing multiple data sources, which creates compartmentalization of information and lack of shared standards. This situation is sometimes worsened by close minded attitude ("data is mine, and I manage it"). DAF tackles this with a 'carrot and stick' strategy. On the one hand, the PA should be willing to use the DAF because it offers evident benefits in terms of efficiency and cost reduction, as most of the general data related activities are centrally managed once and for all PAs; moreover, the PA will have in one single place access to all data from other PAs it may require for its activities, avoiding the cost of managing bilateral agreements with a potential high number of PAs. On the stick side, the regulatory framework will impose to every PAs the obligation to ingest its data into the DAF.

Another critical aspect of data analytics and big data fields is related to the scarcity of skilled human resources. Mastering Big Data technologies and modern machine learning techniques requires highly specialized skills that are not quite easy to find in the market. This means that having every PA to start up a specialized team of this kind is highly expensive if not impractical. A central "dream team" of Big Data and data science experts may be an effective answer to the problem, as it may provide an efficient way to share these skills between PAs when needed, consulting on general purpose analytics matters, and leaving to the PA the focus to domain specific challenges. Moreover, a setup of this kind promote the emergence of economies of scope and cross pollination of knowledge across PAs.

5 Services Provided by the DAF

The DAF is designed to offer general purpose big data and analytics capabilities as well as targeted solutions to specific problems of public administrations. The latters, can be summarized as: interoperability, open data, data products and crowdsourcing.

To enhance interoperability, DAF has a great focus on data governance procedures, making use of an ecosystem of ontologies and controlled vocabularies, a set of ingestion pipelines that performs "standardization" to incoming data, and a rich set of metadata. Every dataset in DAF is accompanied by metadata that describes the dataset and its internal structure. Among them, the user will associate existing ontology information and controlled vocabularies to the dataset data structure, by means of semantic tags. This tagging system is used to provide a logical way to link datasets containing the same semantic tags, and to build standardization procedures that will check and correct for the right use of associated controlled vocabularies, ensuring that all dataset can be effectively joined together.

PAs' data is an asset that is not of exclusive ownership of the PAs themselves. In fact, when there are no privacy issues, PAs' data are of public domain, and should be released as open data to citizens and enterprises. At the time of writing, this effort is far from being systematic and automated, and most of the time is left to the zeal of this or that employee. By ingesting the source data managed by the PAs, the DAF offers automated procedures to transform the original data into a format that can be made public (eliminating all references to privacy related information), and exposed in a standard and machine readable way via rest API. Finally, since DAF manages all metadata required by international standards of open data catalogs, it can offer open data catalog as a services by exposing a unique web application that can be customized by every PA to be used as its own open data portal, freeing the PA from managing "yet another open data catalog".

Data products are software applications that make use of information extracted from data to provide functionalities to end users. An example of a data application may be a program that takes as input the fiscal code of a taxpayer, and returns a likelihood measure of the fact that the taxpayer is a tax evader. Under the hood, the data application looks up the information related to the taxpayer, plug them in a trained machine learning model to detect tax evasion, and returns the results of the model to the user. The DAF is the ideal environment to design, develop and deploy data applications as it provides a unified repository for a wide range of data, a scalable environment for data science and a reliable and fault tolerant platform for deployment.

Finally, DAF provides tools to promote collaboration in data analytics tasks both across PAs and with citizens interested in using public data.

6 Architectural Design Highlights

The core of the DAF architecture (see Fig. 1) is based on a Hadoop cluster and a Kubernetes cluster. The Hadoop cluster is responsible for fault tolerant and horizontal scalable data management operations, and is made of the typical applications of the Hadoop ecosystem. The Kubernetes cluster is responsible for managing all services that add analytical and operational capabilities to the Hadoop cluster. End user services are exposed via the Dataportal, a web application that allows to navigate the

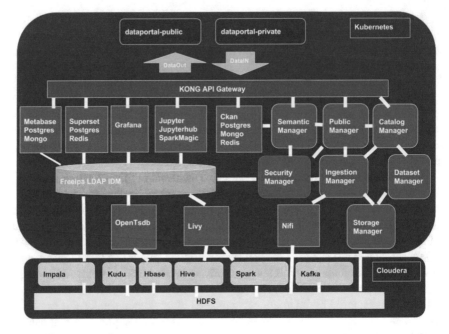

Fig. 1 DAF architecture

data stored in the DAF, and have access to analytic tools for data visualization and data science. Moreover, via the Dataportal, users can share their works with other users of the same organization, or publish them to the world as "data story".

7   Conclusions

New technologies are changing the way data are collected, managed and used to extract information. In fact, Big Data techniques proved to be successful in private sector, helping firms to build competitive advantage against their peers, with measurable impact on their P&L. Public sector is no different: the scale and diversity of information managed makes it a perfect candidate to fully exploit the advantage brought by Big Data techniques. The Data and Analytics Framework is a project promoted by the Digital Transformation Team of the Italian Government, aiming at setting up a central Chief Data Office for the PA. Its main goal is to help PAs to share data among them and with citizens by adopting a centrally managed big data platform to empower them with cutting edge analytics capabilities. Finally, the Framework will support PAs activities with its data experts team that will share its competences and help with analytics projects with an "internal consultancy" fashion.

References

1. David Kiron, Renee Boucher Ferguson and Pamela Kirk Prentice. *On Becoming an Analytical Innovator*. MIT Sloan Management Review, March 2013.
2. Maciejewski, Mariusz. 2017. To do more, better, faster and more cheaply: Using big data in public administration. *International Review of Administrative Sciences* 83: 120–135.
3. McKinsey Global Institute. Big data: The next frontier for innovation, competition and productivity. Strata Summit, September 2011.
4. Blumer, L., C. Giblin, G. Lemermeyer, and J.A. Kwan. 2017. Wisdom within: Unlocking the potential of big data for nursing regulators. *International Nursing Review* 64: 77–82.
5. Giest, Sarah. 2017. *Big Data for Policymaking: Fad or Fasttrack? Policy Sciences* 50 (3): 367–382.
6. Hochtl, J., P. Parycek, and R. Schollhammer. 2016. Big data in the policy cycle: Policy decision making in the digital era. *Journal of Organizational Computing and Electronic Commerce* 26 (1–2): 147–169.
7. McNeely, C.L., and J-o Hahm. 2014. The big (data) bang: Policy, prospects, and challenges. *Review of Policy Research* 31: 304–310. https://doi.org/10.1111/ropr.12082.
8. Mergel, I., R.K. Rethemeyer, and K. Isett. 2016. Big data in public affairs. *Public Admin Rev* 76: 928–937. https://doi.org/10.1111/puar.12625.
9. Stough, R., and D. McBride. 2014. Big data and U.S. public policy. *Review of Policy Research* 31: 339–342. https://doi.org/10.1111/ropr.12083.
10. Yeung, Karen. 2017. Hypernudge: Big data as a mode of regulation by design. *Information, Communication and Society* 20 (1): 118–136.

Advancements in Mobility Data Analysis

Mirco Nanni

Abstract Some recent advancements in the area of Mobility Data Analysis are discussed, a field in which data mining and machine learning methods are applied to infer descriptive patterns and predictive models from digital traces of (human) movement.

Keywords Mobility · Data mining · Trajectory data

1 Big Mobility Data Sources

The current, very fast development of information and telecommunication technologies led an exceptionally wide segment of population to adopt a number of data collection services and information sharing platforms, often based on localization data, such as the family of location-based services (navigation systems, recommendations of local restaurants, etc.). This trend makes it possible, today, to collect information about the mobility of human beings in different contexts, such as the movement of citizens in urban settings or at regional/national geographical scale.

Data sources can be classified into various categories, the most commonly used in scientific literature being the following: vehicle GPS devices, usually adopted for navigational purposes or as black-boxes for insurance clauses; mobile phone data traces (the so called Call Detail Records), which record the antennas where the users placed a call; roadside devices, such as bluetooth/wifi antennas, RFID gates or cameras, which are able to locate an object (usually a vehicle) at given locations in the city; social media data, such as Twitter, Flickr and Instagram, where users that post geo-tagged messages and pictures can be localized in the moment they share a comment or a photo.

Each data source has its own potential information content as well as its own limitations, in terms of spatial accuracy of the localization, sample size, sampling

M. Nanni (✉)
ISTI-CNR, KDDLab, Pisa, Italy
e-mail: mirco.nanni@isti.cnr.it

© Springer International Publishing AG, part of Springer Nature 2018   11
F. Leuzzi and S. Ferilli (eds.), *Traffic Mining Applied to Police
Activities*, Advances in Intelligent Systems and Computing 728,
https://doi.org/10.1007/978-3-319-75608-0_2

rate, privacy issues, etc. In this contribution we focus on one of the most powerful sources, namely vehicle GPS traces, which allow to perform a large range of analyses and infer useful information about human mobility.

In the following sections we summarize some solutions based on the analysis of movements in a collective way, without considering the specific contribution of each user, and then show some approaches that, on the opposite, analyze in depth the individual in order to refine the results. Finally, we describe two sample types of applications and services based on these approaches.

2 Collective Mobility Data Analysis

The raw data provided by GPS devices can be translated into streams of localizations, one for each user, describing where she was at each sample time. In turn, these streams can be divided into single *trips* – or *trajectories* –, each describing a movement between two stop locations. The whole set of trips identified across all users in a geographical area can be used to infer various kinds of summaries, useful to understand its overall mobility (and traffic) [3].

Basic analyses. As simple instance of the mobility indicators that can be derived, we can see when (during the day or week) and where urban mobility is denser, what is the average speed of each road segment, or study the origins of the traffic w.r.t. a set of candidate areas. Similarly, we can study statistical distributions of lengths, durations, speeds, etc. of the trips.

Trajectory clustering. A more sophisticated study consists in discovering the common routes in the dataset, for instance to understand which are the access patterns to a geographical area, such as the center of a city or a mall. That would allow, in particular, to focus the analysis on each pattern and better understand its role in the city traffic. While a traditional approach based on pre-defined origin-destination areas is possible, trajectory data allows a much more flexible solution, where relevant areas directly emerge from the data.

Technically speaking, this can be achieved by clustering trajectories [2], i.e. grouping them into clusters of homogeneous elements: the trajectories within the same cluster are similar to each, while trajectories belonging to different clusters are very different. A first key ingredient to define is a proper notion of similarity between trajectories. This depends on the spacific application at hand. For instance, if we want to identify trajectories with common origins and/or destinations (which is our running example), it is sufficient to compare the starting points and/or the ending points of the trajectories; if we are interested in the specific roads followed to move, then we need to compare the geometrical shape of the trajectories; finally, if we want to find groups of vehicles that moved together, we need to compare the shape and the timing of the trajectories. Several clustering schema are available in literature, yet various works proved that density-based methods (mainly DBSCAN and its variants) are good candidates, since mobility data often show clusters of possibly irregular shape and large proportions of noise.

3 Individual Mobility Data Analysis

The examples discussed above are based on the analysis of trajectories without exploiting their association to an individual, i.e. the fact that several trips form the personal history of a user, which can therefore be understood in better detail.

Mobility profiles. A particular, and very useful, aspect that can be inferred from the set of trips belonging to a single user is the presence of recurrent behaviors, i.e. routines that the user repeats approximately unchanged everyday or so [7]. Technically, routine trips can be discovered by a trajectory clustering analysis performed on the individual history: clusters of significant size will represent recurrent trips, while the rest will form her irregular mobility. We call *Mobility Profile* of a user her set of routines (one for each cluster found). In turn, routines constitute the *systematic* mobility of the user, while irregular trips are her *non-systematic* mobility.

A simple application of mobility profiles is based on the resulting classification of trips into systematic versus non-systematic, which allows to evaluate urban traffic in terms of impact of systematic trips. In our example above, about access patterns to a city center, that allows to distinguish between patterns mainly followed by commuters (the major source of systematic mobility) and patterns preferred by non-systematic travelers.

Mobility prediction. Predicting how the mobility of a vehicle will develop (i.e. which roads it will visit in completing the ongoing trip) is a basic need for forecasting larger scale phenomena, like traffic jams, or for providing timely location-based services to the vehicle (re-routing in case of accidents on the predicted route, as well as commercial offers in gas stations, etc.).

A simple way to predict mobility is to leverage the mobility profiles discussed above [6]. First, if an ongoing trip of a user matches closely with one of her routines, it is very likely that the user will complete the same kind of trajectory. Second, the routines of the other users, although not representative of the personal habits of our user, represent sensible ways for continuing the ongoing trip, capturing customary routes for crossing the city, and therefore a similar matching and prediction process can be applied. This hybrid approach – first try to match the user's routines, then try with those provided by the *crowd* – proved to be very effective in urban settings, enabling several possible applications. Later in this paper, we will show an example in real-time traffic monitoring.

Activity recognition. Systematic and non-systematic trips represent a rather basic classification of users' mobility. A natural evolution of this line is the classification into more complex and refined activities, such as those adopted in transport engineering: going to work, going home, shopping, leisure, bring-and-get (e.g. bringing a child to school), etc.

A (conceptually) simple way to do that consists in checking what kind of activities are actually possible in the geographical area where the vehicle stopped. Indeed, by checking which services and commercial businesses are available within walking distance from the stop location and have operation hours compatible with the arrival time, we could spot several activities. In particular, restaurants, bars and similar can identify *food* activities, while schools and similar can identify (mainly) bring-and-get cases. Beside the possible uncertainty that such methods can involve (e.g. several services might be available in the same place), a limitation is the fact that some activities are more difficult to associate to specific places or commercial businesses.

A more flexible way consists in analyzing the whole mobility of a user by identifying the locations she visited and connecting origin-destination pairs corresponding to each trip, forming a network representation of her mobility history [5]. This global view of the individual allows to characterize each trip based on its relation with the others. In particular, standard mobility features (length and duration of the trip, time of the day it happened, etc.) can be integrated by network-related measures, such as the centrality of the trip within the network, its degree (how many places are connected with the destination location), the predictability of the next trip (basically an entropy measure), etc. A machine learning approach was developed, based on such twofold set of features, which proved to be able to identify activities with very good accuracy, not only for relatively simple classes (like *home* and *work*, usually well identified by the most frequent routines), but also for difficult ones, like *bring-and-get* and others.

Knowing the activity associated to each trip is very useful for better understanding mobility, since now all the analyses mentioned in previous sections can be enriched with an additional semantic layer.

4 Mobility Data-Driven Applications and Services

The sample analyses described so far naturally support several applications. The following are two representative examples in different domains.

Traffic monitoring. When vehicle trajectory data is available in quasi-real time, the traffic measures discussed at the beginning (density in time and space) can be naturally replicated into an online monitoring service. A particular example was developed withing the EU FP7 PETRA project [1], in collaboration with the RSM mobility agency of Rome, Italy. The domain expert selected key areas of the city to monitor (choice that, however, can be dynamically adjusted later), and the system computed, for each area, a simple time series of the traffic (number of trajectories that traversed it) in the last hour, with two important addenda: first, exploiting mobility profiles, the time series also show the percentage of systematic traffic on each area; second, exploiting the trajectory prediction tool, a short-time prediction is performed about the traffic in each area in the next 10 min. Notice that changing the areas to monitor simply implies the re-computation of a few statistics (basically the past values of the time series), which is much easier than approaches based on roadside sensors, where pieces of hardware needs to be physically moved and installed.

Self-organizing mobility: carpooling. Carpooling means sharing a trip with another traveler that, otherwise, would have used her own car for the trip. While several carpooling services have been proposed in the last decade, including online platforms, only very few of them managed to really take off. One reason for that is that the process is always driven by manually-provided requests of the users, which have to plan in advance the possibility of carpooling – either as a driver (looking for passengers) or as a passenger (looking for a lift).

Mobility profiles can be very helpful in that. Indeed, they not only represent habits of the user, but also a mobility demand that she will most likely have also the next day. This can be the basis for a proactive service, that: (i) analyzes the trajectory data of all the users that subscribed to it; (ii) infers the systematic trips of each user, and use them as predicted trip requests for the next day(s); (iii) matches the trips of the users, looking for pairs that can share the travel – notice that they need not to have the same start and end point, but just that one can pick-up and later drop-off the other; (iv) drivers and passengers are paired, trying to globally maximize the number of carpooling users; (v) the carpooling proposals obtained are submitted to the users.

Simulations of the above service [4], performed on real trajectory data covering the Tuscany region, in Italy, showed that the carpooling potential is usually rather high – though variable from area to area –, theoretically allowing around 40% of users to get a lift from other drivers (not considering users that had no routine in the area). Also, the exploration allows to understand where the pick-ups are concentrated (in the experiment, mainly covering large cities and small ones along a few connecting roads).

5   Conclusions

Data collection and localization technology enable several potentially useful analyses and services for the public good, especially in terms of improved urban mobility.

One key point of the process, however, is accessing a (statistically) significant source of data: very few examples of public trajectory datasets are available, due to clear privacy concerns, mainly limited to public transport (taxis) or ad hoc experiments with few volunteers (e.g. GeoLife project and Open Street Map sources). While some scientific research initiatives were carried out on such type of data (also thanks to the tight privacy safety measures that such context can provide), developing working services or large scale studies requires strong efforts in granting access to appropriate data sets.

Also, since trajectories are a complex type of data, sophisticated analyses require to develop ad hoc solutions, often resulting from an adaptation of standard data mining and machine learning methods.

Finally, we remark that understanding mobility can be the first step for applications in several domains, including police activities: traffic control (e.g. the online traffic monitoring described above, the what-if simulation analyses that can be carried out on top of the inferred mobility models, and the deeper understanding of the causes of specific phenomena such as traffic jams in key areas), detecting disruptions (e.g. by detecting mass deviations from routinary trips in specific areas, as symptoms of issues on the roads), fighting crime (e.g. through the analysis of anomalous movements, especially if systematic), etc.

References

1. PETRA: Personal transport advisor: an integrated platform of mobility patterns for Smart Cities to enable demand-adaptive transportation systems. http://petraproject.eu/.
2. Andrienko, G., N. Andrienko, S. Rinzivillo, M. Nanni, D. Pedreschi, and F. Giannotti. 2009. *Interactive Visual Clustering of Large Collections of Trajectories.* VAST: Symposium on Visual Analytics Science and Technology.
3. Giannotti, Fosca, Mirco Nanni, Dino Pedreschi, Fabio Pinelli, Chiara Renso, Salvatore Rinzivillo, and Roberto Trasarti. 2011. Unveiling the complexity of human mobility by querying and mining massive trajectory data. *The VLDB Journal* 20 (5): 695–719.
4. Guidotti, R., M. Nanni, S. Rinzivillo, D. Pedreschi, and F. Giannotti. 2016. Never drive alone: boosting carpooling with network analysis. *Information Systems.*
5. Rinzivillo, Salvatore, Lorenzo Gabrielli, Mirco Nanni, Luca Pappalardo, Dino Pedreschi, and Fosca Giannotti. 2014. The purpose of motion: learning activities from individual mobility networks. In *International Conference on Data Science and Advanced Analytics, DSAA 2014, Shanghai, China, October 30 - November 1, 2014.*
6. Trasarti, R., R. Guidotti, A. Monreale, and F. Giannotti. 2015. Myway: Location prediction via mobility profiling. *Information Systems.*
7. Trasarti, Roberto, Fabio Pinelli, Mirco Nanni, and Fosca Giannotti. 2011. Mining mobility user profiles for car pooling. In *Proceedings of the 17th ACM SIGKDD international conference on Knowledge discovery and data mining, KDD'11,* 1190–1198. New York, NY, USA, ACM.

Part II
Technical Contributions

Towards a Pervasive and Predictive Traffic Police

Fabio Leuzzi, Emiliano Del Signore and Rosanna Ferranti

Abstract The research on traffic flows is historically born to improve road networks, to make trips comfortable and faster. In this research field, as in many others, literary production followed market or business demand. This paper has the objective to clarify police needs, in order to create a research request and to gain attention. It provides an organizational framework concerning needs, goals, fields and some impacts such that different areas of study can concur all together towards a pervasive comprehension of road events, a predictive Traffic Police, so that safety and security can be ensured via a targeted patrolling or intervention. The aim behind the practical level is to pave the way for the interaction between data science from one hand, and law, public administration and justice from the other hand.

Keywords Traffic data mining · Vehicle forensics · Pattern understanding

1 Introduction

Several Police Forces, in Italy, are engaged to improve safety and security. Italian National Police in particular has several sectors in which it is specialist; one of them is the traffic. Among all, it is the only branch of the Police Forces dedicated to the protection of the highways, whereas it is the main reference for the other non-urban motorways, the most important unit in terms of number of men and patrols, and sadly famous for the highest number of losses across all the Police units.

In order to characterize the Italian-Traffic-Police activities, we should underline that it must ensure, as more as possible, the traffic-laws respect, but it has to deal

F. Leuzzi (✉) · E. Del Signore · R. Ferranti
Ministry of the Interior – Italian National Police, Rome, Italy
e-mail: fabio.leuzzi@poliziadistato.it

E. Del Signore
e-mail: emiliano.delsignore@poliziadistato.it

R. Ferranti
e-mail: rosanna.ferranti@interno.it

© Springer International Publishing AG, part of Springer Nature 2018
F. Leuzzi and S. Ferilli (eds.), *Traffic Mining Applied to Police
Activities*, Advances in Intelligent Systems and Computing 728,
https://doi.org/10.1007/978-3-319-75608-0_3

also with accidents (in particular when anyone dies or is critically injured), crimes and highways protection as critical infrastructure.

The spread of electronic technology boosted the production of data in all fields of knowledge without a shared plan or vision. This brought to several sources that made data quickly growing, heterogeneous, time and quality varying, and are so much that cannot be faced with standard techniques of analytics.

Clearly, this technological evolution had an impact in vehicles and traffic as well, leading to a tipping point for the Police Forces, that must enhance the competences in order to keep the control of the public safety and security.

To introduce a bit deeper the Traffic-Police point of view, let us start to list some needs for each branch of competence.

In order to ensure the respect of traffic laws, the Traffic Police will need the automatic check of: vehicle direction, speed, insurance coverage, periodic review, and the correctness of any other maneuver.

For accidents it needs to deeply know the technologies of the newest vehicles, in order to be able to examine the dynamic without hypotheses based on other hypotheses.

To face crimes, it will need Machine Learning and Data Mining approaches able to suggest interesting cases to examine, that could be tied to crimes already done or that could be related to the preparation phase of crimes that can be predicted.

Last but not least, critical infrastructure protection must be taken into account, since highways and vehicles present some security weaknesses; on this latter topic, below is presented a fictional scenario, together with some real hacking events.

Research on traffic flows is historically born to improve road networks, to make trips comfortable and faster. In this research field, as in many others, literary production followed market or business demand. This paper has the objective to clarify police needs, in order to create a research request and to gain attention.

The remainder of the paper starts overviewing open issues in each branch of competence, from the characteristics of the available data to the novelties about vehicle-log recovery and analytics. Then it proposes an integrated approach to the complex issue of traffic and vehicle mining and understanding, with the aim to provide a framework regarding fields of action, impacts and their relationships. The work ends with some final remarks.

2 Research Fields: Background and Challenges

This Section provides background and motivations about the four main research fields (hopefully) composing the next future of the Traffic Police. For every field some related works are cited, together with a synthetic overview of the open issues on which research efforts could focus to practically facilitate police activities.

2.1 Mining Traffic Data

In last years, data sources as phones produced a huge amount of Floating Car Data (FCD). Such a type of data has been broadly used to suggest the best routing to avoid traffic jams. Anyway, for the Traffic-Police purposes, the information that someone is travelling on a given roadway is not enough, since the unit of analysis is the vehicle; in this early stage of evolution, the main required data is the plate number.

The Italian territory is partially covered by plate detection systems, since they are purchased and installed by several type of organizations, that could be governmental, Police Forces or road operators. These systems are not homogeneously distributed on national road network, bringing to a series of consequences. In first place, there is no data continuity on the itinerary of a single vehicle, thus making car flows harder to analyse as a global phenomenon. Secondly, different organizations have different funds that can be used for support and maintenance, implying a different degree of efficiency for each system. Thirdly, for the Italian privacy law, sensible data like transits must be sent to a Police Force, which is the only public subject designed to their storage; unfortunately, several organizations are not compliant with the law, and even if the system is efficient, it does not contribute to the national transits collection system.

Several millions of sensors are distributed over the national road network and several millions of transits are recorded every day, making such data cumbersome as well as quickly growing. As can be easily imagined, systems for transits recognition and storage seem to be very promising in support of police activities, thus several new systems are purchased and installed every year. Such a growing has a twofold consequence, on one hand the Police is unable to predict the growing percentage of the data production per year, given that new systems are installed via free initiatives; on the other hand, this setting makes the Police passive to changes, since it can only wait for new systems, that step by step, will cover the national territory without a shared plan. Furthermore, plate detection cameras have an elevate failure rate, caused by electricity peaks, animals, weather, etc., making the data noisy and/or missing. Another thing to mention is that organizations could vary camera positions, or their total number, making data sources non homogeneous over time. These characteristics bring the challenge up to a big-data level.

Lastly, the Police is not a private company. It is a governmental organization, without any business-oriented strategy, so often it must manage sensible data without the possibility to ask the citizen's agreement. For this reason, being compliant with the privacy laws becomes fundamental. Such laws, from a practical point of view, add difficulties to the data-analytics activities. In fact they introduce restrictions on the maximum data retention time. In other words, some sources of data are treated as temporary streams (like those within transits collection system). This represents a further issue in the field.

As presented in [1], differently from data in traditional static databases, data streams are continuous, unbounded, usually come with high speed and have a data distribution which may change over time. As pointed out in [2], in these cases there is neither enough time to rescan the whole stream each time an update occurs nor enough space to store the entire data stream for on-line processing. Furthermore, the temporal evolution which typically characterizes the distribution of data in a stream demands for techniques that can capture the evolution of extracted patterns as well. A panoramic view about issues and techniques is provided in [3], in which streams are faced at every level, from classical statistics to Machine Learning and Data Mining approaches fitted to work in time windows from which discover knowledge and perform novelty detection.

Although FCD cannot be the center of the analysis, their integration as further features could open novel possibilities to tackle problems unassailable just with plate numbers. Associating moving phones with cars could suggest the number of potential victims in case of crash, or could provide statistics about the ratio of people per vehicle useful for traffic flows predictions, or could allow the estimation of the amount of people that need goods of first necessity to be distributed for aids in case of a huge congestions for a natural disaster, and so on and so forth.

On the wave of traffic laws, a consideration must be dedicated to autonomous cars, since they will open, in the next future, a plethora of challenges, from legal to practical points of view. Italian traffic laws are not ready to receive such a novelty, since responsibilities must be clarified for any type of situation that could happen on the road. Anyway, the practical point of view stimulates more interest in this place, since autonomous cars are totally controlled via software taking decisions on the base of data streams generated by tens of sensors.

As shown in [4], less than 5% of these vehicles can dissipate stop-and-go waves. Such an implication is the tangible proof of the accelerating evolution of traffic flows, producing further challenges and explosions of the amounts of road data.

2.2   Hints of Vehicle Forensics and Analytics

Vehicle forensics is a new branch of digital forensics encompassing the recovery and investigation of data from automotive systems. Nowadays cars are getting more and more sophisticated and consist of a great number of computer components called Electronic Control Units (ECUs), containing a lot of information that can be useful for police activities and so, the need of vehicle forensics experts, will become much more important in the forthcoming future.

It could be very useful to merge and analyse coherently all these kind of information coming from the different electronic systems of the car, in order to investigate an accident involving vehicles (heavy and light), to define the dynamic of a crash or, more generally, to find an illicit event or behaviour.

A first way to extract data from ECUs is using vehicle diagnostic software in combination with a communication interface. Main problems are related to the fact that every manufacturer has its own interface and personalization: there is not a standard defining exactly quantity, quality and format of diagnostic data.

Although useful information can be stored in every ECU and system (for example some information could also be stored in the electronic key) of a vehicle, in a modern car a lot of data are stored in the infotainment system, in the telematics system and in the Event Data Recorder (EDR) module.

In the next Sections, we will see hints of the methods actually available for data recovery from these systems, and related challenges. It is important to note that automotive infotainment and telematics systems must not be confused with EDRs.

Event Data Recorder EDRs cover many different types of devices. The object of this study concerns the subset of devices installed in a vehicle to record technical information for a brief period of time before, during and after a crash.[1] For instance, EDRs may record pre-crash vehicle dynamics, systems status, driver inputs and post-crash data such as the activation of an automatic collision notification system.

The adoption of the EDR system was proposed in 1998 by the National Highway Traffic Safety Agency and, nowadays, vehicles sold in Canada and USA are equipped with it. An American government requirement defines not only what kind of data has to be stored but also its format.[2] Obviously, quantity and quality of data depends on the EDR type.

To the best of our knowledge, actually only one commercial tool in the market is available to read EDR information. The retrieval of data can be done by connecting a computer with the proprietary software to the car (via a hardware interface as OBD) or to the ECU where the EDR is inserted (typically the Airbag Control Module).

European manufacturers have no constraints with the use of this system and a great number of modern vehicles are equipped with it. Differently from Canada and USA, there is not a law reporting specifications about the minimum set of EDRs data that must be ensured for extraction. To overcome this lack, since European and American vehicles have similar EDR modules, it is possible to spoof the EDR system attempting to download data from a different module than intended.[3] As presented in [5], it is feasible to read data from a vehicle that is not officially supported by using a suitable and supported World Manufacturer Identifier or Vehicle Identification Number.

Vehicle Infotainment and Telematics Systems Vehicle infotainment and telematics systems are very common technologies, often confused, as they both overlap in several areas (for instance in the use of the same display and user interface); frequently, even the analysts cannot agree on what exactly differentiates telematics and infotainment. In fact, some manufactures have combined these two systems as one, referring to it as multimedia system.[4]

Vehicle infotainment systems combine software and hardware to offer entertainment and information features as GPS navigation, music streaming, SMS, Bluetooth

---

[1] https://www.nhtsa.gov/research-data/event-data-recorder.

[2] http://www.studiodelcesta.com/2016/01/30/cdr-bosch-panoramica-sul-sistema-e-applicazioni-nella-ricostruzione-dei-sinistri.

[3] http://www.cdr-italia.it/.

[4] https://vin.dataonesoftware.com/vin_basics_blog/vehicle-infotainment-vs-telematics-systems-what-is-the-difference.

connectivity, internet access via Wi-Fi, and so on. While infotainment indicates the combination of information and entertainment, telematics refers to the combination of telecommunications and informatics.[5] An important difference is that a vehicle telematics system has, generally, a two-way communication (to send, receive and store information), it does not include entertainment features and many of the functions run automatically, without an user input. Utilities commonly found in a vehicle telematics system include remote access, notification of vehicle collision, vehicle location by GPS, control of vehicle speed, emergency call, vehicle diagnostics and maintenance notifications.[6]

The infotainment and telematics systems in modern vehicles retain information that could be useful in police activities: event data from cellular telephones (call logs, contact lists, SMS messages, emails, pictures, videos, social media feeds) and other devices connected to the system, vehicle events (timestamp and GPS coordinates of some events like vehicle lights turning on, doors opening and closing, Bluetooth devices connection) and navigation data (recent destinations, favourite locations, navigation history, speed and GPS coordinates).[7]

Actually, we have to face some practical problems and limitations. Today, only one third-party system is available to extract data from the infotainment and telematics system, allowing investigators to access, preserve and analyse vehicle data. Furthermore, due to the lack of standards or regulatory information, these data are not easy to acquire.

Sometimes they can be retrieved via OBD port or connecting physically to the infotainment/telematics module's printed-circuit board (as a common ECU). In some other cases, they can be easily obtained via USB. Unfortunately, several vehicles are not supported yet.

Heavy Vehicles and Tachographs A *tachograph* is a device installed on heavy vehicles in order to automatically record its speed and distance, together with the driver's activity.[8] It can be analog or digital and consists of a sender unit mounted to the vehicle gearbox, a tachograph head and a recording medium. All relevant heavy vehicles manufactured in the EU since May 1st, 2006 must be fitted with the digital version, which has two recording mediums: an internal memory and a digital driver card.

Professional drivers are legally required to accurately record their activities, retain records and produce them on demand to authorities; for this reason, tachographs have a printer that allows to view only the main information about driver's activities, his behaviour (speed, hours of driving, etc.) and system technical data.

Main tachograph manufacturers provide a software to make a less superficial checks of data and logs stored in these systems; furthermore, some software compa-

---

[5]http://www.electronicdesign.com/4g/line-between-telematics-and-infotainment-blurs-even-further.

[6]https://www.extremetech.com/extreme/201026-what-is-vehicle-telematics.

[7]https://digital-forensics.sans.org/blog/2017/05/01/digital-forensics-automotive-infotainment-and-telematics-systems-2.

[8]https://en.wikipedia.org/wiki/Tachograph.

nies develop tools to carry out a deeper examination, analysing data coherency. The need to check logs is becoming crucial because some logistics companies tamper tachograph systems in order to violate the regulations regarding respect for driving and rest time, pushing drivers to work much more than they owe and making them a threat. For this reason, alterations widespread and sophistication are growing to the point of forging consistent data. Furthermore, since logs could be altered, a first challenge relies on the identification of new technologies and techniques to easily detect tachographs tampering at electric/electronical and/or mechanical level.

Smart tachograph [6] is an evolution of the digital one, that should be connected to a positioning service based on a satellite navigation system, recording the position of the heavy vehicle during the daily working period. The cited regulation introduces also the possibility to remotely exchange data with patrols (art. 9); in order to facilitate targeted roadside checks by the competent control authorities. The wireless communication shall be established with the tachograph only when it is requested by the equipment of the control authorities. Data exchange shall be limited to the data necessary for the purpose of targeted roadside checks of vehicles with a potentially manipulated or misused tachograph. An accurate selection of these information will be necessary in order to verify the whole system integrity at all levels. This is a futuristic scenario, especially with respect to the forthcoming Platooning and self-guided cars era.

2.3 Mining Patrolling Data

Almost every patrol engaged in safety and security on the road is equipped with an on-board device aimed to help the policemen to check cars and passengers efficiently, when needed. The specific domain implies noisy and missing data due to the conformation of the territory (often highways pass through mountains and tunnels, then the positioning could be strongly approximated or even absent), as well as totally missing data because such devices are not always turned on.

In any case, even these data are part of the national richness, and could be faced using Spacial Data Mining. As clearly explained in [7], spacial objects are characterized by a geometry which is formulated by means of a reference system. This geometry implicitly defines both spacial properties, such as orientation, and spatial relationships of different natures, such as topological distance or direction relations.

Studies in spacial data structures [8], spacial reasoning [9] and computational geometry [10] have paved the way for the investigation of Spacial Data Mining, which is related to the extraction of the interesting and useful but implicit spacial patterns [11]. A spacial pattern expresses a spacial relationship among (spacial) objects and can take different forms, such as classification rules, association rules, regression models, clusters and trends. Therefore, to extract spatial patterns from spatial datasets the relevant spatial objects, within their properties and relationships must be identified. It is straightforward that this makes the Spacial Data Mining different from traditional Data Mining, since in this last one objects described into the dataset are typically treated as independent observations.

The integration of such patterns with the spatial data about other event can produce an aware resource management.

2.4 Mining Information Exchange Among Control Rooms

To improve the information exchange among control rooms and to address the patrols itinerary better than today, a deep understanding of such exchange is needed so that workflows can be made explicit. As pointed out in [12], while most information for setting up workflow models comes from practitioners, that day by day carry out and improve the procedures, formal models of activities are typically produced by supervisors and managers. This gap often causes models not to perfectly fit the practice. Furthermore, producing models is inherently complex, costly and error-prone [13].

The case of Traffic-Police control rooms surely suffers of this gap, bringing to dangerous degree of freedom in the procedures. Unfortunately, the analysis of particular actions performed by operators is hampered by the difficulties of practitioners to generalize and abstract the procedures.

Hence, we need to study such processes automatically. Let us clarify some basic concepts about this field (well summarized in [12]) so that a line of action to approach to a deeper understanding can be hypothesized. A *workflow* is a formal specification of how set of actions can be composed to result in valid processes. A *case* is a particular execution of actions according to a given workflow. Case *traces* can be obtained by standard software supporting an organization's activity, in the form of lists of events associated to *steps*. Several traces might be collected in *logs*, that interleave their elements [14]. *Process mining* [15] aims at using a set of case traces to infer a formal model of the process behavior [16].

The use of First-Order Logic to represent the knowledge learned by means of the techniques recalled above is referred as *Declarative Process Mining*, recognized as being very important when dealing with particularly complex models and domains [17]. Instead of completely specifying a process flow, it imposes only a set of constraints that must be satisfied when executing the process activities.

The framework proposed in [12], named *WoMan*, works in Inductive Logic Programming adopting the *Datalog* formalism.

Such proposals are particularly relevant to meet our needs, since the logic formalism, satisfying the postulate of comprehensibility [18], makes understandable, then manipulable, the learned processes. This is the main reason for which this field of research seems to be well suited to approach to control-rooms data.

3 An Integrated Approach to Road Understanding and Event Management

This Section provides an organizational framework to clarify needs, goals, fields and some impacts, such that different areas of study can concur all together towards a pervasive comprehension of the road events and a predictive Traffic Police, so that safety and security can be ensured via a targeted patrolling or intervention.

As reported in Fig. 1, the center of any analysis is the event, that could be a traffic jam, an accident, a crime or a natural disaster. Every type of such event can be faced either from inside or outside the car, sometimes both. The "inside" activities can regard logs and connectivity analytics, approaching to forensics, vehicle theft and other crimes, terrorist attacks and violations of traffic laws. The "outside" activities can regard mining traffic data, mining geo-spatial data about patrolling, re-engineering of informative flows among control rooms, approaching to optimization of resources and time of response, monitoring and optimizing software to support informative exchange, understanding of traffic flows to predict them or to detect outliers, and last but not least, emergency response.

As can be seen in Fig. 1, the four main action directions have non-empty intersections about those police activities that will find benefits with scientific results. Such efforts are the base for an understanding of events from several points of view, that in turn is the base for a better management of the road network through predictions in the short period.

Such a scenario is obviously the base for studying and modelling road events for longer predictions, that can allow the detection of deviations and anomalies with respect to the expected values and trends. A deep knowledge and a mature awareness provide the know-how to examine unexpected events with extreme precision, recovering details about dynamics never seen before that will change definitely the way in which flows can be influenced and forensic reports can be brought to the Court.

Crime Contrast Spacing from the traffic mining aimed to detect anomalies to the understanding of the innovative techniques, the contrast of crime can be very various. Several words have been spent above about the need of data mining and analytics, then it seems more useful to provide some details about the comprehension of the on-board technologies that can be tampered, substituted or inserted for tracking or spying, hacked for remote control, removed and/or analysed for log recovery aiming to determine the last movements or events of a vehicle.

As presented in [19], infotainment and telematics data could be beneficial to investigators and accident reconstructionists by:

- linking a particular vehicle to a certain location at a specific date and time;
- identifying the general itinerary of a vehicle for a specific time window;
- linking a particular vehicle to a sequence in question;
- establishing a general time line of events for a period of time under investigation;
- linking a given person to a vehicle at a given date and time;
- identifying potential sources for driver inattention;

Fig. 1 Impacts distribution among the fields of action

- identifying the contacts and potential networks for people under investigation;
- identifying door opening and closing events to determine if someone comes in or out from a particular vehicle at a specific date and time.

An integrated analysis of data coming from ECUs, EDRs, vehicle infotainment and telematics systems and other electronic devices of driver and passengers (smartphones, satellite navigation systems, etc.), even if not explicitly associated to the vehicle, could be central to support the reconstruction of a crash or to identify an illicit behaviour (for example driving behaviour, real mileage of the vehicle, etc.).

Italian Traffic Police aims to prevent and contrast all the crimes related to the road but now we will focus only on tech crimes made on vehicles (in-car objects theft, cars theft, alteration of the vehicle real mileage).

In the recent years, we have seen a significant rise in the use and availability of jamming devices. Jamming of signals is illegal in most countries but identifying an active jammer could be difficult. These devices are used to defeat surveillance and security equipment including cellular devices, GPS and tracking system in order to:

- interfere with the closing door command sent by the keyless entry system (the key of the car usually work at 315 MHz, 433 MHz or 868 MHz) so the car remains open; at this point, the criminal is able to steal objects into the car or take the car itself;
- defeat both commercial on-board navigation and recovery systems, typically jamming on GPS and cellular bands;
- interfere with the anti-theft tracking system.

In all these cases, the availability of a radio-jammer detector, hopefully able to provide a rough localization of these devices, could be very useful.

The business related to the car theft is growing quickly, both in USA and in Europe. In the last years, car thieves improved their techniques and the traditional based methods has been (and will be) progressively substituted by high-tech devices, easy to buy and to use. Some of the main high-tech-thief methods are[9]:

- direct programming of a new key, accessing to the on-board diagnostic port, without having availability of the original key;
- electronic components replacement, mostly of the immobilizer ECU;
- immobilizer override (some systems, once connected to the OBD port, deactivate the immobilizer ECU);
- original key/transponder cloning (it is important to note that this method, unlike the direct key programming, assumes the availability of the original key);
- relay attacks on smart key [20];
- connected-vehicle attacks (in the forthcoming future).

Actually, with the in-use thief method, the attacker must be near or in the proximity of the car; in the future, car thief and much more heavy attacks, could be done remotely. World's major vehicle manufacturers are finally taking the security of vehicles seriously and realizing that what they sell is just a big computer the customer sits in.[10] This topic will be recalled below.

Mortal Accident and Logs A little less than two years ago the legislator, according to the needs expressed by Italian people, aggravated the consequences of an accident if it involves a violation of the traffic laws and anyone dies or is seriously injured. Before this law, the accidents were faced by the police as a scenario in which must be taken measurements in order to solve an administrative controversy. Conversely, today such an accident implies the opening of a criminal proceeding, bringing measurements at a new level, since they are often the key of the judgement, together with expert reports about them. The policeman cannot use ex-post measures to clarify the dynamics, but, in the era of on-board devices, geo-reference and inertial platforms, the policeman must deal with technology, evincing from it the dynamic of impact. As presented in [19], data retrieved from EDRs have been used for years to reconstruct crashes worldwide, especially in USA and Canada. In 2016, the authors of [21], providing a

---

[9]http://www.autotheftblog.it/dossier-sui-furti-dauto-le-nuove-modalita-hi-tech/.

[10]https://www.theguardian.com/technology/2016/aug/28/car-hacking-future-self-driving-security.

thorough review and analysis of all the known validation testing to date, concluded that EDR systems provide valid and useful data that can be used as a supplement to a thorough accident reconstruction. Furthermore, the list of vehicles equipped with EDR, that are compatible with the tool for data recovery as well, is growing day by day.

In the majority of collision events, there is no evidence of brake signs on the road and so, the work of the reconstruction can be very hard and based on hypothesis deriving from vehicle(s) damages. In the pre-crash phase, an EDR reader can extract the speed of the vehicle, pressure on the gas pedal, brake-pedal position, cruise-control state, seat-belt state, steering-wheel position.[11]

In the crash phase, an EDR reader can store directly data like acceleration on the axis, deltaV[12] and duration of the impact.

Vehicle Forensics is a new field and so both in Italy and in other States, the retrieval of data from EDRs, infotainment and telematics systems has still not been fully adopted, specially from an integrated point of view.

Summing up, the EDR system has some limitations (number of supported vehicles, validation of the extracted data) but the retrieval of the EDR data could be crucial to support the reconstruction of a crash, especially if combined and analysed together with data coming from other systems as vehicle and telematics system and other electronic devices (smartphones, satellite navigation systems, etc.) owned by the driver and/or by the other passengers.

Resources Optimization In the last decade, Italy is fully engaged in several spending reviews to optimize resources and maximize the profits for the national system. The Police Forces are quite involved in such reviews, losing men and means every year. The only way to contrast this poorness is optimization, to cite the motto of the current Italian-National-Police chief "do more with less".

From this the need to define the synergy between the Spatial Data Mining on patrolling data and the Process Mining applied to information flows among control rooms, to make even more efficient the patrol response and to improve the proximity to citizens.

Terrorism, a Fictional Scenario The general car system is organized as a composition of interacting applications, this condition opens criticisms about software-errors tolerance, and about cyber security, that can be seen as a threat for the single vehicle, that, by remote control, can be used as a weapon against the passengers or against other people. Furthermore, it opens several scenarios of terrorist threats, one upon all, regarding the highways as critical infrastructure.

Let us imagine a particular model of car having a proper connectivity. It has not the requirement to be autonomous, an independent connectivity is enough. Assume also that the ECUs are connected each other (as today really is for a lot of circulating cars of large distribution). Now imagine that a hacker, in a very hot season, takes the control of all cars of that model (at the same time or sequentially, it does not care),

---

[11] http://www.cdr-italia.it/applicazioni/moti_pre_urto.html.

[12] http://www.cdr-italia.it/applicazioni/analisi_collisione.html.

Without CAN

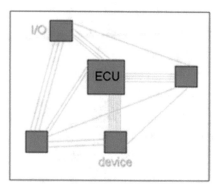

With CAN

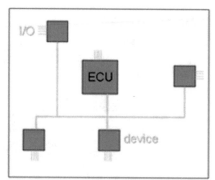

Fig. 2 Digital I/O channels on a modern car

and abrupt imposes the brake to each car. The scenario seems to be apocalyptic: accidents randomly distributed along the highways, unexpected traffic jams, senior citizens locked into the jam, as well as pregnant women, children without water, cars without conditioned air and widespread panic. Emergency call centres constantly busy, unable to sort the emergencies for gravity, ambulances and fire fighters, few in number, all busy and slowed down from chaotic jams.

Such a scenario is just a hint of possible threat that could be avoided studying holes and eventual backdoors in car systems.

In the last few years, some hackers have shown the feasibility of these kind of attacks. In the history of car hacking, we can distinguish three phases[13][14]:

1. Controller Area Network (a.k.a. CAN or CAN-bus) hacking on traditional non-connected car (2011);
2. telematics hacking on connected car (2015);
3. automatic system hacking on autonomous car (2016).

Vehicle hacking is mainly based on the CAN-bus weakness: as shown in Fig. 2,[15] it is a bus born in order to reduce the amount of wiring needed in the vehicle to connect the various electrical components and to provide the communication interface among controllers and sensors on which the OBD protocol relies. The CAN architecture was designed to be light and robust, unfortunately it contains numerous vulnerabilities, as the lack of segmentation and boundary defence, no device authentication and non-encrypted traffic.

---

[13] https://cansecwest.com/slides/2017/CSW2017_MinruiYan-JianhaoLiu_A_visualization_tool_for_evaluating_CAN-bus_cybersecurity.pdf.

[14] https://www.sans.org/reading-room/whitepapers/internet/developments-car-hacking-36607.

[15] https://www.linkedin.com/pulse/using-can-bus-serial-communications-space-flight-christian-mayer.

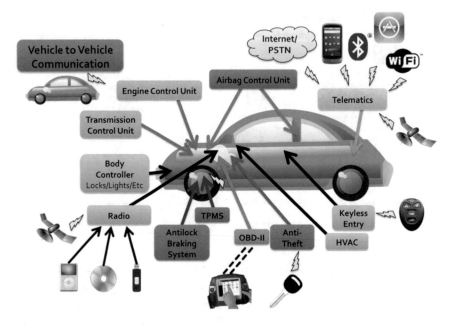

Fig. 3 Digital I/O channels on a modern car

In 2010, the authors of [22] shown that, having wired access to the CAN, it is possible to manipulate vehicle's functions by injecting messages on the bus. In 2011, the same team [23] analysed the external attack surface of a modern car showing that every I/O channel (such as vehicular cellular, Bluetooth, Wi-Fi, etc.) is a potential entry point for an attacker. Figure 3 highlights where digital I/O channels exist on a modern vehicle. In 2013, Miller and Valasek[16] realized a physical hack of a Ford Escape and a Toyota Prius having physical access to the car's BUS and manipulating the behaviour of the vehicle. In 2015, Miller and Valasek[17] realized a remote hack of a Jeep Cherokee, successfully demonstrating that a vehicle could be violated without the need of any physical access, being able to gain remote access and to remotely execute code. In 2016, the authors of [24] examined the security of the sensors of autonomous vehicles. In particular, they presented contactless attacks on these sensors and showed the results collected both in the lab and outdoors on a Tesla Model S. They showed that using off-the-shelf hardware, they were able to perform jamming and spoofing attacks, causing blindness and malfunction, all of which could potentially lead to crashes and impair the safety of self-driving cars. They also proposed software and hardware countermeasures that could improve sensor resilience against these attacks. Recently, to facilitate the process of testing automotive cybersecurity,

---

[16]https://www.forbes.com/sites/andygreenberg/2013/07/24/hackers-reveal-nasty-new-car-attacks-with-me-behind-the-wheel-video/.

[17]https://www.wired.com/2015/07/hackers-remotely-kill-jeep-highway/.

Liu and Yan[18] developed a tool that evaluates the cybersecurity of the CAN-bus, which can be used for black-box tests by security researchers and automotive engineers.

Traffic-Law Violations Engine-ECU tampering (a.k.a. chip tuning) consists in modifying some engine performance parameters in order to achieve boost in power and torque, having a better driving pleasure and, hopefully, reducing fuel consumption. This activity arose at the beginning of 2000 together with the massive spread of turbo-charged diesel engine. Besides reliability problems, it represents a traffic-law violation since the vehicle is not compliant with the original approval (in terms of power, torque, emission, etc.) and so it should be the object of a new technical approval. Reading data from ECUs (see Sect. 2.2) could be a good way to detect this kind of traffic law-violations. Furthermore, the availability of a device or, more generally, of a system able to identify engine-ECU tampering, would be very desirable.

4 Conclusions

Nowadays the road infrastructure is overwhelmed of sensors able to collect data about traffic. Such streams are rapidly stored, and often have all the properties of big data, and more. On the other side, the Traffic Police does always all what can be done, and more, in order to face the dangers and pitfalls that road events imply.

Data analytics, Machine Learning, Data Mining applied to big data and streams will be the tipping point for the Traffic Police, in terms of effectiveness, efficiency, reactivity, awareness about conditions of intervention and safety of operators.

Clearly, in the Big-Data scenario a valuable ethics is needful to evaluate opportunely event predictions, so that underestimations about involved juridical goods can be avoided. Freedom of movement, protection of private-data and data-retention systems, definition of a trade-off among the protected good, motivations of data recording, modalities of access by third parties, data-retention time, rights to data confidentiality or deletion, these are the juridical goods and fundamental rights involved. The legal world is studying and searching solutions about these issues [25, 26] and many others, preparing the legal infrastructure of a State in which humans and thinking machines live together, a futurist scenario in which also the road becomes a secure place, entrusting its ubiquitous safety surveillance to machines, algorithms and mathematical models.

This paper aims to give an overview about police activities on the road, practical difficulties, available data divided by fields of action, open issues about their exploitation, possible impacts that could arise extracting information previously unknown and potentially useful, with the hope to clarify a little bit more needs and directions that research could follow to pursue social order and wellness.

---

[18]https://cansecwest.com/slides/2017/CSW2017_MinruiYan-JianhaoLiu_A_visualization_tool_for_evaluating_CAN-bus_cybersecurity.pdf.

Going further the practical level, we have on one hand data science, and on the other hand law, public administration and justice; the desire has been to pave the way for their deep interaction. To reach such a social evolution, it is equally important to underline the need of a contribution of the legislator, in order to define novel rules to give more degree of freedom to data scientists, making possible predictions without violate the sacredness of sensible data. In the meanwhile, the challenge is to use as better as can be done the margins of available legal freedom.

References

1. Guha, Sudipto, Nick Koudas, and Kyuseok Shim. 2001. Data-streams and histograms. In *Proceedings of the Thirty-third Annual ACM Symposium on Theory of Computing, STOC '01*, 471–475, New York, NY, USA. ACM.
2. Appice, Annalisa, Michelangelo Ceci, Corrado Loglisci, Costantina Caruso, Fabio Fumarola, Michele Todaro, and Donato Malerba. 2009. A relational approach to novelty detection in data streams. In *Proceedings of the Seventeenth Italian Symposium on Advanced Database Systems, SEBD 2009*, 89–100, June 21–24, Camogli, Italy.
3. Gama, Joao. 2010. *Knowledge Discovery from Data Streams*, 1st ed. Chapman & Hall/CRC.
4. Stern, Raphael E., Shumo Cui, Maria Laura Delle Monache, Rahul Bhadani, Matt Bunting, Miles Churchill, Nathaniel Hamilton, R'mani Haulcy, Hannah Pohlmann, Fangyu Wu, Benedetto Piccoli, Benjamin Seibold, Jonathan Sprinkle, and Daniel B. Work. 2017. Dissipation of stop-and-go waves via control of autonomous vehicles: Field experiments. *CoRR*. abs/1705.01693.
5. Singleton, Nathan, Jeremy Daily, and Gavin Manes. 2008. *Automobile Event Data Recorder Forensics*, 261–272. Boston: Springer.
6. European Parliament and Council. 2014. Regulation 165/2014 on tachographs in road transport. In *European Regulation*, 02 2014.
7. Malerba, Donato. 2007. Mining spatial data: Opportunities and challenges of a relational approach. In *IASC 2007*, August 30th September 1st, Aveiro, Portugal.
8. Güting, Ralf Hartmut. 1994. An introduction to spatial database systems. *The VLDB Journal* 3 (4): 357–399.
9. Egenhofer, Max J., and Robert D. Franzosa. 1991. Point-set topological spatial relations. *International Journal of Geographical Information Systems* 5 (2): 161–174.
10. Preparata, Franco P., and Michael I. Shamos. 1985. *Computational Geometry: An Introduction*. New York: Springer.
11. Koperski, Krzysztof, Junas Adhikary, and Jiawei Han. 1996. Spatial data mining: Progress and challenges - survey paper. In *SIGMOD Workshop on Research Issues on data Mining and Knowledge Discovery (DMKD*, 1–10).
12. Ferilli, Stefano. 2014. Woman: Logic-based workflow learning and management. *IEEE Transactions on Systems, Man, and Cybernetics: Systems* 44 (6): 744–756.
13. Herbst, Joachim. 1999. An inductive approach to the acquisition and adaptation of workflow models. In *Proceedings of the IJCAI'99 Workshop on Intelligent Workflow and Process Management: The New Frontier for AI in Business*, 52–57.
14. van der Aalst, W., T. Weijters, and L. Maruster. 2004. Workflow mining: Discovering process models from event logs. *IEEE Transactions on Knowledge and Data Engineering* 16 (9): 1128–1142.
15. Weijters, A.J.M.M., and W.M.P. van der Aalst. 2003. Rediscovering workflow models from event-based data using little thumb. *Integr. Comput. Aided Eng.* 10 (2): 151–162.
16. Cook, Jonathan E., and Alexander L. Wolf. 1998. Discovering models of software processes from event-based data. *ACM Trans. Softw. Eng. Methodol.* 7 (3): 215–249.

17. Pesic, M., and W.M.P. van der Aalst. 2006. A declarative approach for flexible business processes management. In *Proceedings of the 2006 International Conference on Business Process Management Workshops, BPM'06*, 169–180. Berlin: Springer.
18. Michalski, Ryszard S. 1983. *A Theory and Methodology of Inductive Learning*, 83–134. Berlin: Springer.
19. Bortles, William, Sean McDonough, Connor Smith, and Michael Stogsdill. 2017. An introduction to the forensic acquisition of passenger vehicle infotainment and telematics systems data. In *SAE Technical Paper*, 03 2017. SAE International.
20. Garcia, Flavio D., David Oswald, Timo Kasper, and Pierre Pavlidès. 2016. Lock it and still lose it—on the (in)security of automotive remote keyless entry systems. In *25th USENIX Security Symposium (USENIX Security 16)*, Austin, TX. USENIX Association.
21. Bortles, William, Wayne Biever, Neal Carter, and Connor Smith. 2016. A compendium of passenger vehicle event data recorder literature and analysis of validation studies. In *SAE Technical Paper*, 04 2016. SAE International.
22. Koscher, Karl, Alexei Czeskis, Franziska Roesner, Shwetak Patel, Tadayoshi Kohno, Stephen Checkoway, Damon McCoy, Brian Kantor, Danny Anderson, Hovav Shacham, and Stefan Savage. 2010. Experimental security analysis of a modern automobile. In *Proceedings of the 2010 IEEE Symposium on Security and Privacy, SP'10*, 447–462, Washington, DC, USA. IEEE Computer Society.
23. Checkoway, Stephen, Damon McCoy, Brian Kantor, Danny Anderson, Hovav Shacham, Stefan Savage, Karl Koscher, Alexei Czeskis, Franziska Roesner, and Tadayoshi Kohno. 2011. Comprehensive experimental analyses of automotive attack surfaces. In *Proceedings of the 20th USENIX Conference on Security, SEC'11*, Berkeley, CA, USA. USENIX Association.
24. Liu, Jianhao, Chen Yan, and Wenyuan Xu. 2016. Can you trust autonomous vehicles: Contactless attacks against sensors of self-driving vehicle. In *Proceedings of the 24th DEF CON Conference on Hacking*.
25. European Parliament and Council. 1981. Convention for the protection of individuals with regard to automatic processing of personal data. In *European Regulation*, 01 1981.
26. European Parliament and Council. 2017. Guidelines on the protection of individuals with regard to the processing of personal data in a world of big data. In *T-PD(2017)701*, 01 2017.

A Process Mining Approach to the Identification of Normal and Suspect Traffic Behavior

Stefano Ferilli and Domenico Redavid

Abstract Born and typically exploited in the business and industrial application domain, automatic process management, and in particular process mining, might be profitably applied also to the very different domain of traffic understanding. In facts, previous successful experiences in other movement-oriented applications have been reported in the literature. However, some peculiarities of these special domains require powerful techniques to be available. For this reason, these experience exploit the WoMan framework for workflow management, that has proved to be able to handle complex processes. This paper describes the WoMan framework along with its features and functionality, explains why it is more suitable than other process mining approaches and systems available in the current literature, and proposes a number of ways in which it might be applied to the traffic understanding domain. It also highlights possible shortcomings of the WoMan system, that might need adjustments before it can be applied at full scale on real-world traffic data.

Keywords Traffic understanding · Process mining · Activity prediction · Process prediction

1 Introduction

Being able to capture a model of the movements of entities is of utmost importance, especially in the current era, in which georeferencing and mobile technologies are pervasive and allow one to both know the static position of relevant objects, and

S. Ferilli (✉)
Dipartimento di Informatica, Università di Bari, Bari, Italy
e-mail: stefano.ferilli@uniba.it

S. Ferilli
Centro Interdipartimentale per la Logica e sue Applicazioni, Università di Bari, Bari, Italy

D. Redavid
Artificial Brain S.r.l., Bari, Italy
e-mail: redavid@abrain.it

© Springer International Publishing AG, part of Springer Nature 2018
F. Leuzzi and S. Ferilli (eds.), *Traffic Mining Applied to Police Activities*, Advances in Intelligent Systems and Computing 728,
https://doi.org/10.1007/978-3-319-75608-0_4

track the movements of moving entities. For instance, having a model of a person's movements allows to set up appropriate automatic systems that can support that person in carrying out his activities, provide position-based recommendations, and the like.

Traffic is no exception. Being able to model (some aspects of) vehicle traffic on the roads may be crucial for a number of activities and objectives. For instance, one might want to predict traffic jams, or accidents, in order to place appropriate actions aimed at avoiding such events. Or, one might want to identify abnormal behavior of vehicles, that might be associated to suspect activities. In other cases, one might want to exploit the model to improve the check plans and/or the environment in which traffic takes place, removing criticalities and optimizing the overall behavior. We will call *traffic understanding* the general task of obtaining a model of traffic on a given road or in a given region with specific purposes.

We claim that one possible approach to traffic understanding is using Process Management techniques. A *process* consists of actions performed by agents (humans or artifacts) [3, 4]. A *workflow* is a formal specification of how these actions can be composed to result in valid processes. Allowed compositional schemes include sequential, parallel, conditional, or iterative execution [1]. A process execution can be described in terms of *events*, i.e. identifiable, instantaneous actions (including decisions upon the next activity to be performed). A *case* is a particular execution of actions compliant to a given workflow. Case *traces* consist of lists of events associated to *steps* (time points) [2]. A *task* is a generic piece of work, defined to be executed for many cases of the same type. An *activity* is the actual execution of a task by a *resource* (an agent that can carry it out). Relevant events are the start and end of process executions, or of activities [4].

Process management techniques were originally developed to support the business and industrial domain, where the activities of a production process must be monitored and checked for compliance with a desired behavior. If a formal model of the desired behavior is available, process enactment supervision may be enforced, provided that the events related to the process enactment can be detected and delivered to the supervisor. Also, given an intermediate status of a process execution, knowing how the execution will proceed might allow the supervisor to take suitable actions that facilitate the next activities. However, when the domain is complex, manually building the process models that are to be used for supervision, compliance checking, and prediction is very complex, costly, and error-prone. Hence, the need and interest in automatic approaches to carry out the above activities. The solution of automatically learning these models from examples of actual execution, is the task of Process Mining [11, 16].

Since in an industrial environment the rules that determine how the process must be carried out are quite strict, the emphasis is usually more on conformance checking, while prediction of process evolution is more trivial. The emphasis may change significantly if we move toward other, less traditional application domains to which process management can be applied. This paper proposes to apply process mining and management techniques to the task of traffic understanding. In particular, it proposes to use the WoMan framework for workflow management [7], due to its features and to its good performance on similar domains. It is organized as follows. The next two sections provide an overview of the WoMan framework's formalism and functionality. Then, the motivations and proposed approach to apply it to traffic understanding are discussed in Sect. 4. Finally, in the last section, we draw some conclusions and outline future work issues.

2 The WoMan Framework

The WoMan framework [6, 7] lies at the intersection between *Declarative* Process Mining [15] and Inductive Logic Programming (ILP) [13]. It pervasively uses First-Order Logic (FOL for short) as a representation formalism, that provides a great expressiveness potential and allows one to describe contextual information using relationships. Experiments proved that it is able to handle efficiently and effectively very complex processes, thanks to its powerful representation formalism and process handling operators. In the following, we briefly and intuitively recall its fundamental notions, focusing on a user's perspective, rather than on a technical and algorithmic one.

2.1 Input Formalism

WoMan representations [9] are based on the Logic Programming formalism. Specifically, it works in Datalog, where only constants or variables are allowed as terms. Following foundational literature [3, 10], trace elements in WoMan are 7-tuples, represented as facts

$$\text{entry}(T, E, W, P, A, O, R).$$

that report information about relevant events for the case they refer to:

1. T is the event timestamp (all events in a case must have different timestamps),
2. E is the type of the event (one of **begin_process**, **end_process**, **begin_activity**, **end_activity**, and **context_description**),
3. W is the name of the workflow the process refers to,
4. P is a unique identifier for each process execution,
5. A is the name of the activity,
6. O is the progressive number of occurrence of that activity in that process,
7. R (optional) specifies the agent that carries out activity A.

Activity begin and end events allow to properly handle time span and to identify concurrency in task execution, avoiding the need for inferring it by means of statistical (possibly wrong) considerations [2]. When $E = $ **context_description**, A is used to describe contextual information at time T, in the form of a conjunction of FOL atoms built on domain-specific predicates.

2.2 Output Formalism

Given a set of training cases $\mathscr{C}$, WoMan learns a model consisting of a set of atoms built on several predicates, each expressing a different kind of constraint. The fundamental predicates are the following:

- task(t, C_t): task t occurred in training cases C_t.
- transition(I, O, t, C_t): transition[1] t, occurred in training cases C_t, is enabled if all input tasks in $I = [i_1, \ldots, i_n]$ are active; if fired, after stopping the execution of all tasks in I (in any order), the execution of all output tasks in $O = [o_1, \ldots, o_m]$ is started (again, in any order). If several instances of a task can be active at the same time, I and O are multisets, and application of a transition consists in closing as many instances of active tasks as specified in I and in opening as many activations of new tasks as specified in O.

The core of the model, carrying the information about the flow of activities during process execution, is the set of transitions. A convenient notation for expressing transitions is

$$t : I \Rightarrow O \ [C_t]$$

where the C_t parameter can be omitted if irrelevant. A transition $t : I \Rightarrow O$, where I and O are multisets of tasks, is enabled if all input tasks in I are active; it occurs when, after stopping (in any order) the concurrent execution of all tasks in I, the concurrent execution of all output tasks in O is started (again, in any order). For analogy with the notions of 'token' and 'marking' in Petri Nets, during a process enactment we call a *token* an activity that has terminated and can be used to fire a transition, and a *marking* the set of current tokens.

transition/4 atoms express the allowed connections between activities in a very modular way. The transition/4 formalism was proven to be more powerful [6, 7] than Petri or Workflow Nets [16], that are the current standard in Process Mining. It can smoothly express complex task combinations and models involving invisible or duplicate tasks, which are problematic for those formalisms. Indeed, different transitions can combine a given task in different ways with other tasks, or ignore a task when it is not mandatory for a specific passage. Other approaches, by imposing a single component for each task, route on this component all different paths passing from that task, introducing combinations that were never seen in the examples.

Parameter C_t associated to each task or transition t is a multiset because a task or transition may occur several times in the same case, if loops or duplicate tasks

---

[1]Note that this is a different meaning than in Petri Nets.

are present in the model. It plays a fundamental role for several reasons. First, and most important, it allows WoMan to check that all transitions involved in a new execution were all involved in the same (at least one) training case [6]. Second, it allows WoMan to bound the number of repetitions of loops. Indeed, if a task or transition t was executed k times in case c, then C_t includes k occurrences of c, and so WoMan knows the maximum number of times that t can be carried out in the same case. Third, it allows WoMan to compute the probability of a task or transition t, as the relative frequency $|C_t|/n$ where $n = |\mathscr{C}|$ is the number of training cases. This can be used for process simulation, for activity prediction and for noise handling. In particular, as regards the latter, imposing a noise tolerance $N \in [0, 1]$ will cause all tasks/transitions for which $|C_t|/n < N$ to be inhibited in the model during its exploitation.

Transitions can be seen as 'consumers' of their input tasks, and 'producers' of their output tasks. In this perspective, the completion of an activity during a case can be seen as the production of a resource, that is to be consumed by some transition. So, another limitation to the possible combinations of transitions is expressed using the following predicate:

- `transition_provider` $([\tau_1, \ldots, \tau_n], t, q)$: transition t, involving input tasks $I = [i_1, \ldots, i_n]$, is enabled provided that each task $i_k \in I, k = 1, \ldots, n$ was 'produced' as an output of transition τ_k, where the τ_k's are placeholders (variables) to be interpreted according to the Object Identity assumption ("terms (even variables) denoted with different symbols must be distinct (i.e., they must refer to different objects)"); several combinations can be allowed, numbered by progressive q, each encountered in cases C_{tq}.

that partitions the input multiset of a transition according to the producers of the activities to be consumed.

Additional constraints concern the agents that may run the activities:

- `task_agent` (t, A): an agent, matching the roles A, can carry out task t.
- `transition_agent` $([A'_1, \ldots, A'_n], [A''_1, \ldots, A''_m], t, C_{tq}, q)$: transition t, involving input tasks $I = [i_1, \ldots, i_n]$ and output tasks $O = [o_1, \ldots, o_m]$, may occur provided that each task $i_k \in I, k = 1, \ldots, n$ is carried out by an agent matching roles A'_k, and that each task $o_j \in O, j = 1, \ldots, m$ is carried out by an agent matching roles A''_j; several combinations can be allowed, numbered by progressive q, each encountered in cases C_{tq}.

WoMan can handle taxonomies of agent roles. Each A'_k or A''_j is an expression in disjunctive normal form:

$$(r_{11} \wedge \cdots \wedge r_{1n_1}) \vee \cdots \vee (r_{m1} \wedge \cdots \wedge r_{mn_m})$$

where each r_{ij} is an individual or a role in the taxonomy, meaning that the agent must match all roles in at least one disjunct. The conjuncts are introduced to handle multiple inheritance. The generalization/specialization relationship is handled, in that a role is considered as matched by an agent if the agent matches any of its subclasses in the taxonomy. During the mining phase, generalizing means replacing one or more roles/instances with one of their superclasses.

Time constraints can be set on WoMan models using the following predicates:

- task_time $(t, [b', b''], [e', e''], d)$: task t must begin at a time $i_b \in [b', b'']$ and end at a time $i_e \in [e', e'']$, and has average duration d;
- transition_time $(t, [b', b''], [e', e''], g, d)$: transition t must begin at a time $i_b \in [b', b'']$ and end at a time $i_e \in [e', e'']$; it has average duration d (from the beginning of the first activity in I to the end of the last activity in O), and requires an average time gap g between the end of the last input task in I and the activation of the first output task in O;
- task_in_transition_time $(t, p, [b', b''], [e', e''], d)$: task t, when run in transition p, must begin at a time $i_b \in [b', b'']$ and end at a time $i_e \in [e', e'']$, and has average duration d;

where i_b, b', b'', i_e, e', and e'' are relative to the start of the process execution, i.e. they are computed as the timestamp difference between the **begin_process** event and the event they refer to.

In addition to the exact timestamp of events, WoMan internally associates each activity in a case to a unique integer identifier, called *step*, assigned by progressive start timestamp. So, the above constraints may be expressed also in terms of steps, as follows:

- task_step $(t, [b', b''], [e', e''], d)$: task t must start at a step $s_b \in [b', b'']$ and end at a step $s_e \in [e', e'']$, along an average number of steps d;
- transition_step $(t, [b', b''], [e', e''], g, d)$: transition t must start at a step $s_b \in [b', b'']$ and end at a step $s_e \in [e', e'']$; it takes place along an average number of steps d (from the step of the first activity in I to the step of the last activity in O), and requires an average gap of g steps between the end of the last input task in I and the beginning of the first output task in O;
- task_in_transition_step $(t, p, [b', b''], [e', e''], d)$: task t, when run in transition p, must start at a step $s_b \in [b', b'']$ and end at a step $s_e \in [e', e'']$, along an average number of steps d;

Finally, WoMan can express pre- and post-conditions for tasks (in general), transitions, and tasks in the context of a specific transition. Specifically, conditions on transitions define when a transition may take place; task conditions define what must be true for a given task in general, task in transition conditions define further constraints for allowing a task to be run in the context of a specific transition (provided that its general conditions are met). They are defined as FOL rules of the following form:

- act_$T(A, S, R)$:- … meaning that "activity A, of type T, can be run by agent R at step S of a case execution provided that …";
- trans_$T(S)$:- … meaning that "transition T can be run at step S of a case execution provided that …";

- `act_T_in_trans_P(A, S, R):-` ... meaning that "activity A, of type T, can be run by agent R in the context of transition P at step S of a case execution provided that ...";

where the premises '...' are conjunctions of atoms based on contextual and control flow information. Conditions are not limited to the current status of execution. They may involve the status at several steps using two predicates:

- `activity(s, t)`: at step s (unique identifier) t is executed;
- `after(s', s",[n',n"],[m', m"])`: step $s"$ follows step s' after a number of steps ranging between n' and $n"$ and after a time ranging between m' and $m"$.

Due to concurrency, predicate `after/3` induces a partial ordering on the set of steps. The difference between pre- and post- conditions is that premises in the former refer only to steps up to S, while in the latter they may refer to any step, both before and after S.

3 Workflow Supervision and Prediction

WoMan's learning module, **WIND** (Workflow INDucer), can learn or refine a process model according to a case, after the case events are completely acquired. The refinement may affect the structure and/or the probabilities. Differently from all previous approaches in the literature, it is *fully incremental*: not only can it refine an existing model according to new cases whenever they become available, it can even start learning from an empty model and a single case, while others need a (large) number of cases to draw significant statistics before learning starts. This is a significant advance with respect to the state-of-the-art, because continuous adaptation of the learned model to the actual practice can be carried out efficiently, effectively and transparently to the users [6].

The supervision module, **WEST** (Workflow Enactment Supervisor and Trainer), can check whether new cases are compliant with a given model. It takes the case events as long as they are available, and returns information about their compliance with the currently available model for the process they refer to. The output for each event can be 'ok', 'error' (e.g., when closing activities that had never begun, or terminating the process while activities are still running), or a set of warnings denoting different kinds of deviations from the model (e.g., unexpected task or transition, preconditions not fulfilled, unexpected resource running a given activity, etc.).

Given a partial process enactment, there may be different combinations of transitions that are compliant with that partial enactment. Each of these combinations determines an alternate *status* of the process. Since one may not know which is the correct one until a later time, all possible alternate statuses must be carried on by the system. Also, given one of such statuses and a new activity that is started, there may be different valid ways in which that event can lead to evolution of that status. On one hand, each of these evolutions is a possible new status of the process enactment, posing again the aforementioned ambiguity issues. On the other hand, the new event may point out that some current alternate statuses were wrong, which causes those statuses to be removed because they have been found out to be inconsistent with later events in the process enactment. So, as long as the process enactment proceeds, the set of alternate statuses that are compliant with the activities carried out so far can be both expanded with new branches, and pruned of all alternatives that become incompatible with the activities carried out so far.

Each alternate status may be compliant with a different set of training cases, and may rise different warnings. WEST takes note of the warnings for each status and carries them on, because they might reflect secondary deviations from the model that one is willing to accept. So, each status records the following information:

- the *marking*, i.e., the set terminated activities that have not been used yet to fire a transition, each associated with the agent that carried it out and to the transition in which it was involved as an output activity;
- the set of activities that are *ready* to start, i.e., the output activities of transitions that have been fired in the status, and that the system is waiting for in order to complete those transitions;
- the set of training cases that are compliant with that status;
- the set of (hypothesized) transitions that have been fired to reach that status;
- the set of warnings raised by the various events that led to that status.

The system also needs to remember, at any moment in time, the set *Running* of currently running activities and the list *Transitions* of transitions actually carried out so far in the case. The set of statuses is maintained by WEST, as long as the events in a case are processed.

While in supervision mode, the prediction modules can be used to foresee which activities the user is likely to perform next, or to understand which process is being carried out among a given set of candidates. Due to the discussed set of alternate statuses that are compliant with the activities carried out at any moment, differently from Petri Nets it is not obvious to determine which are the next activities that will be carried out. Indeed, each status might be associated to different expected evolutions. WoMan leverages the information associated to the set of alternate statuses to compute statistics on the expected activities in the different statuses, and uses these statistics to rank by confidence all possible predictions. Confidence here is not to be interpreted in the mathematical sense. It is determined based on a heuristic combination of several parameters associated with the possible alternate process statuses that are compliant with the current partial process execution.

Specifically, **SNAP** (Suggester of Next Action in Process) hypothesizes which are the possible/appropriate next activities that can be expected given the current intermediate status of a process execution, ranked by confidence. Specifically, the activities that can be carried out next in a given status are those included in the *Ready* component of that status, or those belonging to the output set of transitions that are enabled by the *Marking* component of that status. The status parameters used for the predictions are the following:

1. frequency of activities across the various statuses (activities that appear in more statuses are more likely to be carried out next);
2. number of cases with which each status is compliant (activities expected in the statuses supported by more training cases are more likely to be carried out next);
3. number of warnings raised by the status (activities expected in statuses that raised less warnings are more likely to be carried out next);
4. confidence of the tasks and transitions as computed by the multiset of cases supporting them in the model (activities supported by more confidence, or belonging to transitions that are associated to more confidence, are more likely to be carried out next).

Finally, given a case of an unknown workflow, **WoGue** (Workflow Guesser) returns a ranking (by confidence) of a set of candidate process models. Again, this prediction is based on the possible alternate statuses identified by WEST when applying the events of the current process enactment to the candidate models. In this case, for each model, the candidate models are ranked by decreasing performance of their 'best' status, i.e. the status reporting best performance in (one or a combination of) the above parameters.

4 Proposal for Application to Traffic Understanding

After presenting the WoMan framework, in this section we discuss why and how we propose to apply it to the traffic understanding task.

4.1 Setting

First of all, let us define the elements of the 'traffic' process.

- The process is the typical way of traversing the various gates that are placed along the selected road. There are at least two levels at which the traffic can be modeled: single vehicle or overall traffic. In the following we will adopt the former perspective, focusing on the behavior of a single vehicle.[2] Also, different granularities can be adopted. For the time being, we will consider a single 'compound' process accounting for the whole dataset. Different (and, depending on the objectives, possibly more appropriate) choices might be defining different processes for some relevant combinations of the different periods of the year, days of the week, and hours of the day.

---

[2] By this, we do not mean that we learn the behavior of a *specific* vehicle. While learning the behavior of specific vehicles would be possible and, maybe, interesting, our aim here is learning the behavior of a generic vehicle on the road, based on the observed behavior of all vehicles in the dataset, taken in isolation (i.e., independently of the behavior of other cars that are present on the same road at the same time).

- In the above perspective, the only involved agent is a generic vehicle.
- A case, in the single-vehicle perspective, is a complete traversal of the road by a specific vehicle. Since the same vehicle may traverse (portions of) the selected road several times in a sufficiently wide time span (e.g., commuters might traverse it at least once a day), deciding when a case starts and ends is not obvious. E.g., using midnight as a case delimiter would not work for overnight traversals. We will adopt the simple approach by which the first occurrence of a vehicle starts a case, that ends when the vehicle is not detected anymore for a fixed number of hours. This allows to consider as a single case also a traversal in which the vehicle repeatedly enters and exits the road (which may be a feature of suspicious behavior).
- Activities are the positions of the vehicles as detected by the system, i.e., the gates. More fine-grained activities might be defined encompassing both the gate and the lane in which the vehicle is detected.
- The current dataset does not provide much contextual information, if any. If relevant, the lane and/or the presence of other vehicles might be considered as contextual information. Also, the period of the day/week/year might be considered as contextual information, if not already considered when defining the process models of interest.

4.2   Motivation

The idea of applying WoMan to traffic understanding comes from the useful features it provides to tackle the peculiarities of this task, and from the successful results obtained in the past on similar tasks [8].

As to the features, we think that the following are relevant:

Ability to handle complex processes (involving optional activities and duplicate activities, short and nested loops, high concurrency). In particular, optional activities may be present when a sensing device misses a vehicle, while loops may happen if a vehicle exits the road and then enters it again through a previous gate.

Noise handling Not only this may be used for obtaining a model of the typical routes of vehicles by stripping out infrequent routes. From an opposite perspective, keeping only infrequent routes may represent the set of behaviors among which looking for 'malicious' or dangerous ones.

Time handling Especially in the perspective of identifying suspicious behaviors, checking whether a vehicle's pace lies between the limits of usual pace may be very important. On the contrary, too fast or too slow traversals of the gates are very relevant, also for the objective of determining which segments of the road deserve more attention and control.

Ability to handle contextual information If available, contextual information may be precious to grasp relevant features of the vehicles' behavior. Indeed, some contextual features may determine the behavior of a vehicle.

Flexibility In the traffic domain, quite differently than in industrial processes, there is much more variability and subjectivity in the users' behavior, and there is no 'correct' underlying model, just some kind of 'typicality' can be expected.

Incrementality Normal and abnormal traffic are concepts that may evolve in time, e.g. due to changing regulations, changing driving habits of people, etc. An incremental approach to modeling traffic would allow us to quickly incorporate changes into the model, and thus to keep the model always up-to-date.

Efficiency Efficiency is very important to handle large amounts of data and to be responsive in real-world applications. WoMan proved to be more efficient than competitors at the state of the art, but the size of data expected in real-world traffic understanding applications is expected to be so huge that optimizations will be certainly needed.

To the best of our knowledge, no previous attempt have been made to apply Process Mining approaches to the traffic domain. Compared to other state-of-the-art Process Mining approaches, WoMan provides this unique mix of features that we think are necessary to deal with traffic understanding. Compared to other approaches to traffic understanding, we think that incrementality, the ability to handle complex interactions, and the possibility of obtaining human-readable models are the most important contributions that WoMan can bring to bear.

As to the previous results, WoMan proved to be able to learn useful models in the following domains, that (we think) share several similarities with traffic.

People's moving habits at home Specifically, WoMan was applied to the real-world dataset *GPItaly*, built by extracting the data from one of the Italian use cases of the GiraffPlus Ambient Assisted Living project[3] [5]. It concerns an elderly person in her home, and focuses on the movements of her home's inhabitant(s) in the various rooms of the home. The dataset considered 253 days, each of which was considered as a case of the process representing the typical movements of people in the home. Tasks, here, correspond to rooms, while transitions correspond to leaving a room and entering another. The resource (i.e., the person that moves between the rooms) is always the same.

Chess playing can also be considered somehow similar. The selected dataset consists of 400 reports of actual top-level matches played by Anatolij Karpov and Garry Kasparov (200 matches each) downloaded from the Italian Chess Federation website.[4] In this case, chess pieces move around the chessboard, from squares to squares. Specifically, playing a chess match corresponds to enacting a process, where a task is the occupation of a specific square of the chessboard by a specific kind of piece (e.g., "black rook in a8" denotes the task of occupying the top-leftmost square with a black rook), and transitions correspond to moves: indeed, each move of a player terminates some activities (since it moves pieces away from the squares they currently occupy) and starts new activities (that is, the occupation by pieces of their destination squares). The involved resources are the

---

[3]http://www.giraffplus.eu.
[4]http://scacchi.qnet.it.

Table 1 Dataset statistics

	Cases	Events		Activities		Tasks		Transitions	
		Overall	Avg	Overall	Avg	Overall	Avg	Overall	Avg
GPItaly	253	185844	369.47	92669	366.28	8	0.03	79	0.31
White	158	36768	232.71	18226	115.35	681	4.31	4083	25.84
Black	87	21142	243.01	10484	120.51	663	7.62	3006	34.55
Draw	155	32422	209.17	16056	103.59	658	4.25	3434	22.15

two players: 'white' and 'black'. In this dataset we distinguished three processes, corresponding to the possible match outcomes: *white* wins, *black* wins, or *draw*.

Table 1 reports some statistics on these datasets: number of cases, events, activities, tasks and transitions. The average number of events, activities, tasks, and transitions per case is also reported. There are more cases for the GPItaly dataset than for the chess ones. However, the chess datasets involve many many more different tasks and transitions, many of which are rare or even unique. In facts, the number of tasks and transitions is much less than the number of cases in GPItaly, while the opposite holds for the chess datasets. As regards the number of events and activities, GPItaly is more complex than the chess datasets. The datasets are different also from a qualitative viewpoint. In GPItaly, many short loops, optional and duplicated activities are present, but no concurrency. On the other hand, the chess datasets are characterized by very high parallelism: each game starts with 32 concurrent activities (a number which is beyond the reach of many current process mining systems [7]). This number progressively decreases as long as the game proceeds and pieces are taken, but remains still high (about 10 concurrent activities) even at the end of the game. Short and nested loops, optional and duplicated tasks are present as well.

Activity Prediction For prediction experiments, each dataset was split into training and test sets using a k-fold cross-validation procedure (see column *Folds* in Table 2 for the values of k): for GPItaly, 3 folds used due to the very large number of cases and activities in this dataset; for the others, 5 folds were used to provide the system with more learning information. Then, WIND was used to learn models from the training sets. Finally, each model was used as a reference to call WEST, SNAP and WoGue on each event in the test sets: the former checked compliance of the new event and suitably updated the set of statuses associated to the current case, while the latter used the resulting set of statuses to make predictions.

Table 2 (section 'Activity') reports performance averages concerning the different processes (row 'chess' refers to the average of the chess sub-datasets). Column *Pred* reports in how many cases SNAP returned a prediction. Indeed, when tasks or transitions not present in the model are executed in the current enactment, WoMan assumes a new kind of process is enacted, and avoids making predictions. Among the cases in which WoMan makes a prediction, column *Recall* reports in how many of those predictions the correct activity (i.e., the activity that is actually carried out next) is present. Finally, column *Rank* reports how close it is to the first element of

Table 2 Prediction statistics

	Folds	Activity						Process						
		Pred	Recall	Rank	(Tasks)	Quality	Pos	(%)	C	A	U	W
GPItaly	3	0.99	0.97	0.96	8.02	0.92	—	—	—	—	—	—
Black	5	0.42	0.98	1.0	11.8	0.42	2.06	(0.47)	0.20	0.00	0.15	0.66
White	5	0.55	0.97	1.0	11.27	0.54	1.60	(0.70)	0.44	0.00	0.15	0.40
Draw	5	0.64	0.98	1.0	10.95	0.62	1.78	(0.61)	0.29	0.01	0.18	0.52
Chess	5	0.54	0.98	1.0	11.34	0.53	1.81	(0.59)	0.31	0.00	0.16	0.53

the ranking (1.0 meaning it is the first in the ranking, possibly with other activities, and 0.0 meaning it is the last in the ranking), and *Tasks* is the average length of the ranking. The *Quality* index is a mix of these values, obtained as

$$Quality = Pred \cdot Recall \cdot Rank \in [0, 1]$$

It is useful to have an immediate indication of the overall performance in activity prediction. When it is 0, it means that predictions are completely unreliable; when it is 1, it means that WoMan always makes a prediction, and that such a prediction is correct (i.e., the correct activity is at the top of the ranking).

First, note that, when WoMan makes a prediction, it is extremely reliable. The correct activity that will be performed next is almost always present in the ranking (97–98% of the times), and always in the top section (first 10% items) of the ranking. For the chess processes it is always at the very top. This shows that WoMan is effective under very different conditions as regards the complexity of the models to be handled. The number of predictions is proportional to the number of tasks and transitions in the model. This was expected, because, the more variability in behaviors, the more likely it is that the test sets contain behaviors that were not present in the training sets. WoMan is almost always able to make a prediction in the Ambient Intelligence domain, which is extremely important in order to provide continuous support to the users. The percentage of predictions drops significantly in the chess domain, where however it still covers more than half of the match. Interestingly, albeit the evaluation metrics are different and not directly comparable, the *Quality* is at the same level or above the state-of-the-art performance obtained using Deep Learning [12] and Neural Networks [14]. The nice thing is that WoMan reaches this percentage by being able to distinguish cases in which it can make an extremely reliable prediction from cases in which it prefers not to make a prediction at all. The worst performance is on 'black', possibly because this dataset includes less training cases.

Process Prediction The process prediction task was evaluated on the chess dataset, because it provided three different kinds of processes based on the same domain. So, we tried to predict the match outcome (white wins, black wins, or draw) as long as match events were provided to the system. The experimental procedure was similar to the procedure used for the activity prediction task. We used the same folds as for the other task, and the same models learned from the training sets. Then, we merged the test sets and proceeded as follows. On each event in the test set, WoGue was called using as candidate models *white*, *black*, and *draw*. In turn, it called WEST on each of these models to check compliance of that event and suitably update the corresponding sets of statuses associated to the current case. Finally, it ranked the models by increasing number of warnings in their 'best' status, where the 'best' status was the status that raised less warnings. Note that, in this case, WoMan always returns a prediction.

Table 2 (section 'Process') summarizes the performance on the process prediction task. Column *Pos* reports the average position of the correct prediction in the ranking (normalized in parentheses to [0, 1], where 1 represents the top of the ranking, and 0 its bottom). The last columns report, on average, for what percentage of the case duration the prediction was correct (C: the correct process was alone at the top of the ranking), approximately correct (A: the correct process shared the top of the ranking with other, but not all, processes), undefined (U: all processes were ranked equal), or wrong (W: the correct process was not at the top of the ranking). All kinds of chess processes show the same behavior. The overall percentage of correct predictions (C) is above 30%, which is a good result, with A being almost null. Studying how the density of the different predictions is distributed along the match, we discovered that, on average, the typical sequence of outcomes is: U-A-C-W. This could be expected: total indecision happens at the beginning of the match (when all possibilities are still open), while wrong predictions are made towards the end (where it is likely that each single match has a unique final, never seen previously). In the middle of the game, where the information compiled in the model is still representative, correct or approximately correct predictions are more dense, which is encouraging. This also explains the high percentage of wrong predictions (53%): since every match becomes somehow unique starting from 1/2–3/4 of the game, the learned model is unable to provide useful predictions in this phase (and indeed, predicting the outcome of a chess match is a really hard task also for humans). So, the percentage of correct predictions should be evaluated only on the middle phase of the match, where it is much higher than what is reported in Table 2.

4.3 Example

For illustration purposes, let us now show what input and output might be expected when applying WoMan to the TRAP-2017 dataset. Let us consider the trip of vehicle with (obfuscated) plate number 2062495 on August 15th, 2016. The corresponding process case would be described by the following events:

```
entry(20160815200107,begin_of_process,trap17,trap17_15082016_2062495,start,0,none).
entry(20160815200107,begin_of_activity,trap17,trap17_15082016_2062495,act_19,1,ag_I).
entry(20160815200428,end_of_activity,trap17,trap17_15082016_2062495,act_19,1,ag_I).
entry(20160815200428,begin_of_activity,trap17,trap17_15082016_2062495,act_25,1,ag_I).
entry(20160815201403,end_of_activity,trap17,trap17_15082016_2062495,act_25,1,ag_I).
entry(20160815201403,begin_of_activity,trap17,trap17_15082016_2062495,act_20,1,ag_I).
entry(20160815202118,end_of_activity,trap17,trap17_15082016_2062495,act_20,1,ag_I).
entry(20160815202118,begin_of_activity,trap17,trap17_15082016_2062495,act_2,1,ag_I).
entry(20160815202856,end_of_activity,trap17,trap17_15082016_2062495,act_2,1,ag_I).
entry(20160815202856,begin_of_activity,trap17,trap17_15082016_2062495,act_16,1,ag_I).
entry(20160815204259,end_of_activity,trap17,trap17_15082016_2062495,act_16,1,ag_I).
entry(20160815204259,begin_of_activity,trap17,trap17_15082016_2062495,act_4,1,ag_I).
entry(20160815204259,end_of_activity,trap17,trap17_15082016_2062495,act_4,1,ag_I).
entry(20160815204259,end_of_process,trap17,trap17_15082016_2062495,stop,0,none).
```

meaning that the vehicle was Italian, it entered the road and passed the follow-ing sequence of gates during its trip: 19 (at 8:01:07pm), 25 (at 8:04:28pm), 20 (at

8:14:03pm), 2 (at 8:21:18pm), 16 (at 8:28:56pm), 4 (at 8:42:59pm) before exiting the road.

Let us now show and comment some fragment of the model learned from transits occurred on August 15th, 2016:

```
...
task(act_2,[3-1,5-1,6-1,8-1,13-1,14-1,29-1,37-1,...]).
task(act_5,[7-1,11-1,19-1,22-1,23-1,25-1,28-1,34-1,36-1,...]).
task(act_7,[4-1,14-1,35-1,48-1,49-1,80-1,91-1,...]).
task(act_14,[10-1,15-1,36-1,42-1,45-1,74-1,87-1,93-1,94-1,95-1,102-1,...1975-2,...]).
task(act_15,[11-1,15-1,19-1,22-1,23-1,25-1,28-1,...]).
task(act_17,[1-1,20-1,30-1,34-1,35-1,43-1,...,820-2,...,1634-2,...]).
task(act_20,[3-1,5-1,8-1,14-1,29-1,39-1,40-1,...]).
...
```

task atoms say that gate 2 was passed once in cases 3, 5, 6, etc.; gate 5 was passed once in cases 7, 11, 19, etc.; and so on. We may see that gate 14 was passed twice in case 1975, and gate 17 was passed twice in two cases (820 and 1634), which might be interesting for analysis purposes.

```
...
transition([act_2],[act_7],[14-1,80-1,190-1,...,1545-2,...],41).
transition([act_5],[act_14],[36-1,45-1,74-1,...,442-1,467-2,...,867-1,870-2,...1528-2,...],63).
transition([start],[act_20],[40-1,121-1,282-1,315-1,...],66).
transition([act_17],[act_7],[49-1,183-1,193-1,...,820-2,...],72).
transition([act_20],[stop],[67-1,117-1,139-1,180-1,267-1,282-1,...],76).
transition([act_20],[act_15],[120-1,326-1,467-2,506-1,...],90).
...
```

transition atoms say that gate 7 was passed next to gate 2 once in cases 14, 80, 190, etc.; twice in case 1545; and so on (transition 41). Transition 63 (passing gate 14 next to gate 5) occurred twice in several cases (467, 867, 1582). Gate 7 was passed sometimes next to gate 2 (transition 41) and sometimes next to gate 17 (transition 72). After gate 20 some cars passed gate 15 (transition 90), some others finished their trip (transition 76). Some cars started their trip from gate 20 (transition 66), some cars finished their trip after gate 20 (transition 76).

```
...
task_step(act_2,[1,14],[1,14],1.0).
task_step(act_5,[1,13],[1,13],1.0).
...
task_time(act_2,[0,58716],[0,59469],618.2).
task_time(act_5,[0,76866],[0,77188],882.1).
...
```

task_step atoms say that: gate 2 was always between the 1st and 14th gate passed by all vehicles; gate 5 was always between the 1st and 13th gate passed by all vehicles; and so on. All gates were passed in just one step (again, because no concurrency is considered in this dataset). task_time atoms say that: gate 2 was always entered between the beginning of the trip and 58716 units of time from the beginning of the trip, and was always exited between the beginning of the trip and 59469 units of time from the beginning of the trip, within a timespan of 618.2 units of time; and so on.

...
transition_step(63,[1,12],[2,13],2.0,0.0).
transition_step(66,[0,0],[1,1],2.0,0.0).
transition_step(76,[1,15],[2,16],2.0,0.0).
transition_step(90,[2,10],[3,11],2.0,0.0).
...
transition_time(63,[333,49591],[333,49591],3601.56,0.0).
transition_time(66,[0,0],[0,23228],1132.11,0.0).
transition_time(76,[0,77430],[0,77430],0.0,0.0).
transition_time(90,[10276,81696],[10488,81696],232.55,0.0).
...

transition_step atoms say that: transition 63 (i.e., passing gate 14 next to gate 15) always started between the 1st and 12th gate passed by all vehicles, and ended between the 2nd and 13th gate passed by all vehicles; transition 66 (i.e., passing gate 20 as the first gate in the trip) always started with the start of the trip and ended at the first gate passed; and so on. All transitions took exactly two steps to be carried out (one for each gate passed). As regards timings, transition_time atoms say that: transition 90 (i.e., passing gate 15 next to gate 20) always started between 10276 and 81696 units of time from the beginning of the trip, it always ended between 10488 and 81696 units of time from the beginning of the trip, and it lasted 232.55 units of time on average; transition 66 (i.e., passing gate 20 as the first gate in the trip) always started exactly at the beginning of the trip, it always ended between the beginning of the trip and 23228 units of time from the beginning of the trip, and it lasted 1132.11 units of time on average; transition 76 (i.e., passing gate 20 as the last gate in the trip) always started and ended between the beginning of the trip and 77430 units of time from the beginning of the trip, lasting 0 units of time (because there was no next gate); and so on.

...
task_in_trans_step(act_7,41,[2,14],[2,14],1.0).
task_in_trans_step(act_14,63,[2,13],[2,13],1.0).
task_in_trans_step(act_20,66,[1,1],[1,1],1.0).
...
task_in_trans_time(act_7,41,[184,55882],[202,75758],3154.35).
task_in_trans_time(act_14,63,[333,49591],[333,49591],3601.56).
task_in_trans_time(act_20,66,[0,0],[0,23228],1132.11).
...

`task_in_trans_step` atoms say that: gate 14, seen in transition 63 (i.e., passed next to gate 5) was always between the 2nd and 13th gate passed by all vehicles (so, it was never the first gate passed); gate 20, seen in transition 66 (i.e., passed as the first gate in the trip), was always exactly the first gate passed; and so on. All gates were passed in just one step (this is because no concurrency is considered in this dataset). `task_in_trans_time` atoms say that: gate 7, when passed next to gate 5 (transition 41), was always entered between 184 and 55882 units of time from the beginning of the trip, and was always exited between 202 and 75758 units of time from the beginning of the trip, and it was passed after 3154.35 units of time from the beginning of the trip on average; gate 20, when passed as the first gate in the trip (transition 66), was always entered immediately at the beginning of the trip, and was always exited between 0 and 23228 units of time from the beginning of the trip, and it was passed after 1132.1 units of time from the beginning of the trip on average; and so on.

```
...
transition_provider([act_5-_],63,1).
transition_provider([act_20-_],76,1).
transition_provider([act_20-_],90,1).
...
```

`transition_provider` atoms say that for all the transitions reported above, any previous transition ('_' meaning 'any') may have provided the previous gate: e.g., any previous transition may have provided gate 5 as the input task of transition 63, any previous transition may have provided gate 20 as the input task of transition 76, and so on.

```
...
task_agent(act_7,[[ag_RO],[ag_none],[ag_I]]).
task_agent(act_17,[[ag_SLO],[ag_RO],[ag_none],[ag_I],[ag_CH]]).
task_agent(act_20,[[ag_HR],[ag_H],[ag_RO],[ag_CH],[ag_none],[ag_I]]).
...
transition_agent([[[ag_none],[ag_I]]],[[[ag_none],[ag_I]]],41,[14-1,80-1,190-1,...],1).
transition_agent([[[ag_HR],[ag_H],[ag_I],[ag_none]]],[[[ag_HR],[ag_H],[ag_I],[ag_none]]],
63,[36-1,45-1,74-1,...,442-1,467-2,...,867-1,870-2,...,1512-1,1528-2,...],1).
transition_agent([[[ag_none],[ag_I]]],[[[ag_none],[ag_I]]],90,[120-1,326-1,467-2,...],1).
...
```

`task_agent` atoms say that all gates were passed by vehicles of unknown nationality or Italian, but in particular: gate 7 was passed also by Romanian vehicles; gate 17 was passed also by Slovenian, Romanian and Swiss vehicles; and so on. `transition_agent` atoms say that all transitions were carried out by vehicles of unknown nationality or Italian, and in particular: transition 41 (i.e., passing gate 7 next to gate 2) was carried out just once per trip in cases 14, 80, 190, etc.; transition 63 (i.e., passing gate 14 next to gate 5) was carried out also by Croatian and Hungarian vehicles, and (interestingly) there are several trips (467, 870, 1528) in which it was carried out twice; and so on.

5 Conclusions and Future Work

While traditionally exploited for checking process enactment conformance, process models may be used to predict which activities will be carried out next, or which of a set of candidate processes is actually being executed. The prediction performance may also provide clues about the correctness and reliability of the model. In some applications, such as Ambient Intelligence ones, there is more flexibility of behaviors than on industrial applications, which makes these predictions both more relevant and harder to be obtained.

This paper proposes a way to apply the WoMan framework for workflow management to the task of traffic understanding, motivated by the interesting features of this framework and by successful experimental results on similar tasks and domains. For instance, using the data gathered from sensors (e.g., cameras) placed along a given road, it might be possible to learn an overall model of a given road, and to use it both for understanding typical ways in which that road is exploited, and uncommon ways of exploiting it (outliers). The former might help the Police to better plan its activities on that road, or to identify critical issues to be solved on that road. The latter might help to spot, also automatically, illegal or otherwise dangerous behaviors of road users.

Of course, the traffic understanding domain has specificities and peculiarities that must be addressed, and that may require changes and adaptation of the WoMan framework, which will be the subject of our future work. One change will almost certainly be needed to deal with the huge amount of data that is expected to be involved in this new domain, compared to previous application of WoMan and of Process Management approaches in general.

Acknowledgements Thanks to Amedeo Cesta and Gabriella Cortellessa for providing the GPItaly dataset, and to Riccardo De Benedictis for translating it into WoMan format. This work was partially funded by the Italian PON 2007–2013 project PON02_00563_3489339 'Puglia @Service'.

References

1. van der Aalst, W. 1998. The application of petri nets to workflow management. *The Journal of Circuits, Systems and Computers* 8: 21–66.
2. van der Aalst, W., T. Weijters, and L. Maruster. 2004. Workflow mining: Discovering process models from event logs. *IEEE Transactions on Knowledge and Data Engineering* 16: 1128–1142.
3. Agrawal, R., D. Gunopulos, and F. Leymann. 1998. *Mining process models from workflow logs*, Proceedings of the 6th International Conference on Extending Database Technology (EDBT).

4. Cook, J., and A. Wolf. 1996. Discovering models of software processes from event-based data. Tech. Rep. CU-CS-819-96, Department of Computer Science, University of Colorado.

5. Coradeschi, S., A. Cesta, G. Cortellessa, L. Coraci, J. Gonzalez, L. Karlsson, F. Furfari, A. Loutfi, A. Orlandini, F. Palumbo, F. Pecora, S. von Rump, A. Štimec, J. Ullberg, and B. tslund. 2013. *Giraffplus: Combining social interaction and long term monitoring for promoting independent living*. Proeedings of the 6th International Conference on Human System Interaction (HSI) 578–585, IEEE.

6. Ferilli, S. 2014. Woman: Logic-based workflow learning and management. *IEEE Transaction on Systems, Man and Cybernetics: Systems* 44: 744–756.

7. Ferilli, S., and F. Esposito. 2013. A logic framework for incremental learning of process models. *Fundamenta Informaticae* 128: 413–443.

8. Ferilli, S., D. Redavid, and S. Angelastro. 2016. Predicting process behavior in woman. In *Advances in Artificial Intelligence: AI*IA2016*, vol. 10037, ed. G. Adorni, S. Cagnoni, M. Gori, and M. Maratea, 308–320., Lecture Notes in Artificial Intelligence Berlin: Springer.

9. Ferilli, S., D. Redavid, and S. Angelastro. 2017. Activity prediction in process management using the woman framework. *Advances in Data Mining. Applications and Theoretical Aspects*, vol. 10357, 194–208., Lecture Notes in Artificial Intelligence Berlin: Springer.

10. Herbst, J., and D. Karagiannis. 1999. *An inductive approach to the acquisition and adaptation of workflow models*, Proceedings of the IJCAI'99 Workshop on Intelligent Workflow and Process Management: The New Frontier for AI in Business, pp. 52–57.

11. 2012. Process mining manifesto. *Business Process Management Workshops*, vol. 99, 169–194., Lecture Notes in Business Information Processing Berlin: Springer.

12. Lai, M. 2015. Giraffe: Using deep reinforcement learning to play chess. CoRR abs/1509.01549 (2015), arXiv:1509.01549

13. Muggleton, S. 1991. Inductive logic programming. *New Generation Computing* 8 (4): 295–318.

14. Oshri, B., and N. Khandwala. Predicting moves in chess using convolutional neural networks. In: Stanford University Course Project Reports – CS231n: Convolutional Neural Networks for Visual Recognition (Winter 2016), cs231n.stanford.edu/reports/ConvChess.pdf

15. Pesic, M., and W.M.P van der Aalst. 2006. *A declarative approach for flexible business processes management*, Proceedings of the 2006 International Conference on Business Process Management Workshops BPM'06, 169–180, Springer.

16. Weijters, A., and W. van der Aalst. 2001. *Rediscovering workflow models from event-based data*, Proceedings of the 11th Dutch-Belgian Conference of Machine Learning (Benelearn 2001), ed. V. Hoste, and G.D. Pauw, pp. 93–100.

Detecting Criminal Behaviour Patterns in Spain and Italy Using Formal Concept Analysis

Jose Manuel Rodriguez-Jimenez

Abstract Automatic number plate reading systems (NPRS) collect considerable amount of information from roads: number of vehicles, movements, legal status, etc. An immense quantity of information does not represent an answer to a problem if we cannot define what we are looking for and cannot extract knowledge from this information. Formal concept analysis is not recommended for big data, but it has interesting tools to extract knowledge from information stored in databases. Pruning consists in reducing initial information, done by discarding a selectable number of data that we consider not relevant. If pruned properly, the size of the database is reduced but interesting information are retained. Considerable resources are required to assess specific criminal behaviour profiles and research can help to determine which profiles we are interested in. In this paper, we focus on observed behaviour patterns in criminal activities committed in Southern Spain to reduce information provided by NPRS on Italian roads. With this reduced information we conclude that a consensus on appropriate data analysis could be reached if we focus on specific profiles.

1 Introduction

Automatic number plate recognition (ANPR), vehicle licence plate recognition (VL-PR) or automatic licence plate recognition (ALPR) are different names for NPRS. It is a technology that uses optical character recognition (OCR) on images to read vehicle registration plates. NPRS are used by Police forces around the world for law enforcement purposes, including to check if a vehicle is registered or licensed. The information provided by these systems is also used for electronic toll collection and

J. M. Rodriguez-Jimenez (✉)
University of Malaga, Andalucia Tech, Malaga, Spain
e-mail: jmrodriguez@ctima.uma.es

J. M. Rodriguez-Jimenez
Mijas Police Department, Mijas, Spain

© Springer International Publishing AG, part of Springer Nature 2018
F. Leuzzi and S. Ferilli (eds.), *Traffic Mining Applied to Police Activities*, Advances in Intelligent Systems and Computing 728,
https://doi.org/10.1007/978-3-319-75608-0_5

as a method of cataloguing the movements of traffic. This information, not designed for law enforcement purposes, could also be used to prevent criminal activities.

NPRS can be used to store the images captured by the cameras as well as the text from the licence plate. If we are only interested in number plate recognition, cameras can take images from front and rear of the vehicle. Front images are not useful if the vehicle is a motorbike and there exists some cameras with configurable options to store a photograph of the driver that cannot be obtained from rear images, so the configuration of these cameras offers different information. It is desirable that systems use infrared lighting to allow the camera to take the picture at any time of the day and NPRS technology must take into account plate variations from different countries. For example, Italian and French plates have the same configuration, 2 letters, 3 numbers and 2 letters, that differs from Spanish plates that are 4 numbers and 3 letters.

The information stored specifies what (not always who), where and when a vehicle is located and this information can be analysed to obtain knowledge about driver behaviours.

The greatest problem that we have to face is that there exists an excess of information that have to be removed. There are thousands of vehicles in public roads every day that generate new registers in databases bringing no added value. Dealing with this information is a very tricky situation and has to be handled properly if we want to obtain the desired results.

Related works about the specific task described in this study, criminal pattern analysis based on vehicles movements, were not found, so we consider how Formal Concept Analysis could be used in this problem. There exists several works about traffic density measures using traffic cameras [1, 12, 13], but are not related with this study. Some patterns with known routes are studied with statistics [4].

In this work, the background is divided in two sections. While in Sect. 2, the theory is introduced in a brief description of Formal Concept Analysis, the practical part is introduced in Sect. 3, where previous studied criminal behaviours detected in Spain are detailed. In Sect. 4, we explore if there are vehicles that has behaviours related with the criminal behaviours detected in Southern Spain by Local Police forces. Conclusions and future work are described in Sect. 5.

2   Formal Concept Analysis

In this section, the basic notions related with Formal Concept Analysis (FCA) [11] and attribute implications are briefly presented. See [2] for a more detailed explanation.

A *formal context* is a triple $\mathbb{K} = \langle G, M, I \rangle$ where G and M are finite non-empty sets and $I \subseteq G \times M$ is a binary relation. The elements in G are named objects, the elements in M attributes and $\langle g, m \rangle \in I$ means that the object g has the attribute m. From this triple, two mappings $\uparrow: 2^G \to 2^M$ and $\downarrow: 2^M \to 2^G$, named derivation operators, are defined as follows: for any $X \subseteq G$ and $Y \subseteq M$,

$$X^\uparrow = \{m \in M \mid \langle g, m \rangle \in I \text{ for all } g \in X\} \tag{1}$$

$$Y^\downarrow = \{g \in G \mid \langle g, m \rangle \in I \text{ for all } m \in Y\} \tag{2}$$

$X^\uparrow$ is the subset of all attributes shared by all the objects in X and $Y^\downarrow$ is the subset of all objects that have the attributes in Y. The pair $(\uparrow, \downarrow)$ constitutes a Galois connection between 2^G and 2^M and, therefore, both compositions are closure operators.

A pair of subsets $\langle X, Y \rangle$ with $X \subseteq G$ and $Y \subseteq M$ such that $X^\uparrow = Y$ and $Y^\downarrow = X$ is named a *formal concept* where X is its *extent* and Y its *intent*. These extents and intents coincide with closed sets wrt the closure operators because $X^{\uparrow\downarrow} = X$ and $Y^{\downarrow\uparrow} = Y$. Thus, the set of all formal concepts is a lattice, named *concept lattice*, with the relation

$$\langle X_1, Y_1 \rangle \le \langle X_2, Y_2 \rangle \text{ if and only if } X_1 \subseteq X_2 \text{ (or equivalently, } Y_2 \subseteq Y_1) \tag{3}$$

This concept lattice will be denoted by $\mathfrak{B}(G, M, I)$.

The concept lattice can be characterized in terms of *attribute implications* being expressions $A \to B$ where $A, B \subseteq M$. An implication $A \to B$ holds in a context $\mathbb{K}$ if $A^\downarrow \subseteq B^\downarrow$. That is, any object that has all the attributes in A has also all the attributes in B. It is well known that the sets of attribute implications that are valid in a context satisfies the so called Armstrong's Axioms:

[Ref]   Reflexivity: If $B \subseteq A$ then $\vdash A \to B$.
[Augm]   Augmentation: $A \to B \vdash A \cup C \to B \cup C$.
[Trans]   Transitivity: $A \to B, B \to C \vdash A \to C$.

A set of implications Σ is considered an *implicational system for* $\mathbb{K}$ if: an implication holds in $\mathbb{K}$ if and only if it can be inferred, by using Armstrong's Axioms, from Σ.

Armstrong's axioms allow to define the closure of attribute sets wrt an implicational system (the closure of a set A is usually denoted as A^+) and it is well-known that closed sets coincide with intents. On the other hand, several kind of implicational systems has been defined in the literature being the most used the so-called Duquenne–Guigues (or stem) basis [3]. This basis satisfies that its cardinality is minimum among all the implicational systems and can be obtained from a context by using the renowned NextClosure Algorithm [2].

A more general framework is necessary to deal with the information provided by the absence of information. In [7], we have tackled this issue focusing on the problem of mining implication with positive and negative attributes from formal contexts. As a conclusion of that work we emphasized the necessity of a full development of an algebraic framework that was initiated in [6]. A more detailed explanation of the algebraic framework can be seen in [9, 10].

For reasons of simplicity, in order to help the reader, formal context are shown in tables, where appears 1 when an object has an attribute and 0 in the opposite case (See an example in Fig. 1). Also, the absence of an attribute is denoted with a line over the attribute.

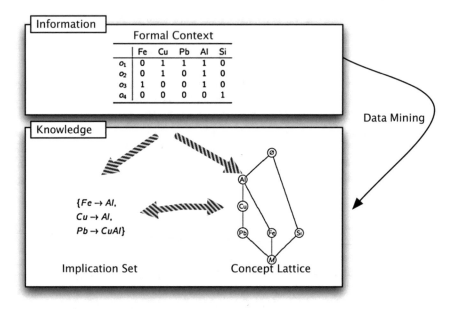

Fig. 1 From information to knowledge in FCA

Since all data sets can be converted to FCA datasets, we can use FCA methods in all possible scenarios, defining properly objects and attributes. This basic idea let us to consider the advantages of using FCA, obtaining knowledge from information provided by data sets (See Fig. 1).

In traffic scenarios, the set of objects are defined as vehicles (or plates) and the different attributes could be defined answering different questions. The answer to questions that have two values (yes or no) define properties of the object (vehicle), for example, was this vehicle registered in a specified day? or, was this vehicle registered by this camera?

With this information, applying FCA methods to obtain basis of implications or concept lattices, we can extract knowledge from the data provided by traffic cameras. Since we have too much not necessary information, FCA methods could reduce it better than statistical methods, considering the relation among different attributes.

3 Criminal Behaviour Patterns in Southern Spain

In Southern Spain there exists three different types of Police forces. National Police and Guardia Civil are national forces with research departments for criminal activities that have different specialized groups. They have full access to national vehicle and criminal databases, EUCARIS and also EUROPOL and INTERPOL databases.

Local Police works in their own cities and only have access to national vehicle and criminal databases. This is the reason why they cannot investigate properly criminal activities although they have direct contact with citizens, and also criminals, in the neighbourhood. They don't have access to traffic or surveillance cameras, but they manage information that can help in analysing criminal behaviour patterns.

This information is improved by sharing data between the nearest Local Police Departments. With this shared information we can know the movements of particular potential criminal groups and prevent criminal offences.

Most cases, criminal groups are conformed by foreign people from Morocco, Romania or Bulgaria. A few years ago, they used foreign cars in an attempt to confound Police forces pretending to be tourists. Most of these vehicles are unlicensed so, with the information provided by international public databases, these vehicles were removed from roads.

Nowadays, the pattern changed and criminals use cars rented with false documents. We cannot know efficiently what vehicles they use because they are changed every two or three weeks, but we can infer their behaviour pattern.

Analysing Police reports we discovered patterns related to external events. They know where and when Police forces are usually located, like temporary markets, and where and when Police forces take a lot of time in coming, so they adapt their schedule to Police schedule. These reports content place and time where thieves act, but we cannot observe patterns until we add the day of the week. Then, with this detail, we can describe all the zones where they are located and the external events that could explain why they are in these zones.

In the other hand, we can infer the relation among criminal groups analysing Police reports with identified potential criminals as was described in [8].

All this data was shared by Local Police forces located in cities from Southern Spain and the knowledge that we obtain turn into the most effective tool to avoid criminal activities.

It was discovered that a few groups move around an extended zone. The live in a place where they don't realize criminal activities and move to cities that are located far from the original place. It is possible that they repeat the same point once or twice every week.

There exists another profile that is denominated "nomad". A group moves around Spain, stay more or less a week in a zone, get all possible money and moves to other zone. Detecting their movements is very difficult because they don't have a clear pattern and a fixed route.

So we have two profiles that we considered are evaluable objectives:

- Vehicles that appears in the same zone once or twice in a week for one month or more.
- Vehicles that appears in the same zone around five days in a week, and then they don't appear again.

4 Analysing Datasets of Traffic Cameras

In this section, we are going to analyse datasets from [5] using the patterns described in previous section and FCA with negative attributes. Since we do not have additional information about reports of criminal activities, we can only point to vehicles that we considered have to be checked and offer a description of their movements.

A computer AMD E2-1800 with Windows 10 and 4 Gb of RAM was used for checking the database with a program that was developed in JAVA.

Due to the described profiles that will be checked, the study could be reduced to a restricted period in time. The considered minimum time could be a month, so in the used examples we will consider December 2016. Extended periods could be used to confirm profiles or discard vehicles previously checked.

An initial formal context was built with the information from datasets. Objects are vehicles (plates) and attributes are days of the year, so there is said that a vehicle was registered in a specified day when, at least one of the cameras from the studied zone, take an image of this vehicle. FCA methods will not be used in this formal context, where we have the necessary information to consider if a vehicle could be a candidate for the first or second profile, or we can discard it, reducing the size of the formal context.

A partial context is shown in Table 1. Vehicles with plates 275417 and 1978769 could be discarded because they are not frequently registered. Vehicles with plates 145318 and 1231984 could be associated with the first profile, but they have a difference between them because the second one is regularly registered. Vehicle with plate 2264772 could be associated with second profile since it is frequently registered, but they are not images after 21st December. Most of registers could be discarded because they could not be associated to the selected profiles.

It expended 15844 ms to get the plates registered in December 2016 and 126956551 ms extracting the profile of movements in the month of the 121656 plates registered on 1st December 2016.

It was checked that the time to get the plates registered in the first fifteen days of December 2016 expend only 7656 ms and the profiles for these days of the 121656 plates registered on 1st December 2016 were extracted in 49304743 ms that is a third part of the time expend in the profiles of all month. This faster solution was discarded because the information provided by these profiles is not enough for our purposes.

Table 1 Formal context from vehicles registered in December 2016

Plate	1	2	3	4	5	6	7	8	9	0	1	2	3	4	5	6	7	8	9	0	1	2	3	4	5	6	7	8	9	0
145318	1	1	0	0	0	1	0	1	0	0	0	0	1	0	0	1	0	0	0	1	1	1	0	0	0	0	1	0	0	0
275417	1	0	0	0	0	0	0	0	0	0	0	0	0	0	0	0	0	0	0	0	0	0	0	0	0	0	0	0	0	0
1231984	1	0	0	0	0	0	0	1	0	0	0	0	0	0	1	0	0	0	0	0	1	0	0	0	0	0	0	0	0	0
1978769	1	0	0	0	0	0	0	0	0	0	0	0	0	0	0	0	0	0	0	0	0	0	0	0	0	0	0	0	0	1
2264772	1	1	1	0	1	1	1	0	1	1	0	1	1	1	1	0	0	0	1	1	1	0	0	0	0	0	0	0	0	0

The elaboration of the profiles could be improved using pruning methods related to the required profile and also plates from vehicles that were discarded in previous analysis could be removed decreasing the size of the database to explore.

A secondary formal context was built if we are interested in the information related to a specified vehicle (plate). Objects are days of the year, that changed the role from the initial formal context, and attributes are numbered cameras, so the information collected from datasets refers to the selected vehicle was registered in a specified day from a numbered camera, at least once.

In this formal context, we can apply FCA methods, considering negative attributes, for extracting knowledge about the behaviour of the vehicle with systems of implications between attributes and concept lattices, but a problem was found. Analysing existing information, there are absence of information in the dataset about vehicles that have to be registered in selected days by a selected camera. The reasons are unknown and could be related to failures in the registration process such an error reading the plate.

Systems of implications between attributes could relate possible routes for the selected vehicles, so a prediction model about the next step for this vehicle could be developed.

Concept lattices could show the structure of visited cameras for each vehicle and classify them by relevance. This classification helps Police forces in their work, pointing to relevant areas.

4.1 First Profile

First profile is related to vehicles that appear once or twice in a week for one month or more. This profile corresponds to criminal groups that move around a central place following different routes. This profile could also be related to vehicles from transport companies, so each case have to be analysed separately.

To describe how we are going to analyse data, examples will be used. As was indicated at the beginning of this section, we will analyse movement of vehicles in December 2016 to check if there exists vehicles with the profile that we want to find. It means vehicles that appear in the dataset between four and eight days in the month and their appearances are separated in time more than three days.

In the secondary context, we apply FCA methods to obtain a system of implications between attributes. With this knowledge we can infer relations among these attributes (vehicle registered by a specified camera). Also, we add a complementary study about speed between cameras to detect irregularities since reports from criminal activities in the area are unknown.

In Table 1 two candidates were found. The vehicle with plate 145318 has not a regular pattern and their appearances are not separated in time more than three days, so it will not be analysed.

The vehicle with plate 1231984 appears on days 1, 8, 15 and 22, i.e. every Thursday except 29th. Considering the route that is covered by cameras 9, 26, 10, 18, 23, 15,

8, 3 and 13 a system of implications between attributes was inferred (camera 5 is excluded because never registered the vehicle).

- $13 \rightarrow 9, \overline{26}, 10, 18, \overline{23}, \overline{15}, 8, \overline{3}$
- $\overline{13} \rightarrow 15$
- $\overline{3} \rightarrow 9, \overline{26}, 18, \overline{23}, 8$
- $8 \rightarrow 18$
- $15, 3, \overline{13} \rightarrow \overline{9}, 26$
- $\overline{23}, 15, \overline{13} \rightarrow \overline{10}$
- $18 \rightarrow 8$
- $18, \overline{23}, 8 \rightarrow 9, \overline{26}, \overline{3}$
- $10 \rightarrow 18, 8$
- $\overline{10}, 15, \overline{13} \rightarrow \overline{23}$
- $\overline{26} \rightarrow 9, 18, \overline{23}, 8, \overline{3}$
- $9 \rightarrow \overline{26}, 18, \overline{23}, 8, \overline{3}$
- $\overline{9}, 26, \overline{10}, \overline{23}, 15, 3, \overline{13} \rightarrow \overline{18}, \overline{8}$
- $9, \overline{26}, 10, 18, \overline{23}, 8, \overline{3} \rightarrow \overline{15}, 3$

This system of implications could be improved fixing the existing problem with intermediate cameras. If it is supposed that the vehicle passed under the intermediate cameras, the system of implications is this new one that focus the knowledge in start and finish points of the route.

- $\emptyset \rightarrow 26, 10, 18, 23, 15, 8$
- $13 \rightarrow 9, 3$
- $3, \overline{13} \rightarrow \overline{9}$
- $\overline{3}, \overline{13} \rightarrow 9$

There exists some facts that could be interesting for this vehicle analysing data shown in Tables 2 and 3:

- It seems that have a regular route from camera 9 to camera 13 but there are absence of registers in some intermediate doors (50, 50, 62.5 and 12.5% respectively). We consider that the percentage of fails is so high. Camera 5 never registered the vehicle, but we can observe that this camera registered other vehicles.
- On days 1st and 15th there exists an individual register in door 22 in the middle of the night. It could be the start of the trip in the opposite way for the route detected, but we cannot affirm this possibility because there is not information to confirm it.
- There is a low speed detected between cameras 9 and 18 in day 1st. This speed is an outlier if we compare with average speed of this vehicle in the same point. There is a service area between these cameras, so we have to study if there exists a problem in road traffic at this time or there exists a report of criminal activity in this service area at this time.

Table 2 Time of registration from vehicle 1231984 in December 2016

Day	9	26	10	18	23	15	5	8	3	13
1	21:55:30			22:23:31		22:40:39		22:52:44		
8	19:44:34		19:56:06	20:02:47				20:24:52		20:32:46
15		18:08:43				18:38:43			18:55:44	
22		21:16:33	21:19:20	21:25:12	21:32:18	21:39:09		21:49:46	21:53:14	

Table 3 Speed calculated from vehicle 1231984 in December 2016

Day	9(31,1)	26(48,5)	10(54,5)	18(66,8)	23(80,8)	15(98,4)	5(106,1)	8(117,8)	3(124,7)	13(134,5)
1	X			76.45		110.66		96.33		
8	X		121.73	128.38				138.57		126.84
15		X				99.8			92.73	
22		X	129.34	146.25	101.41	154.16		109.64	119.42	

4.2 Second Profile

Second profile is related to vehicles that appear four or five times in the same week in the same zone, and they do not appear again. This profile corresponds to "nomads" criminal groups that move around a particular zone for a short time.

In this case, a system of implications of attributes to detect possible routes is not the better option because their movements could be random. Proposed FCA methods infers a concept lattice that has a tree structure in order to show the relevance of the different attributes.

Since systems of implications between attributes and concept lattices have the same knowledge, the selection of the second one for this profile is only for a better understanding of the knowledge represented.

As an example, we are going to analyse movements of vehicle with plate 2264772 in December 2016. In November, this vehicle appears only twice, on day 7th appears in camera 4 and on day 23rd appears registered for camera 3. These registers are separated in space and time so they could be related to movements in zones out of the controlled area that cover 18 numbered cameras.

In Table 4 we can observe the registered movements for this vehicle that points to some interesting facts:

- We infer with a ? in Table 4 that this vehicle have to be registered at least for these cameras.
- On initial days of December, it moved over a specific zone but changed to other area from day 9th to 21st, not appearing after this date.
- Camera 16 checked six times this vehicle, so we consider that criminal reports in rest areas near this point have to be analysed. This vehicle has three lonely registers in this point so it could be a reference to Police forces.

Table 4 Registration for vehicle 2264772 in December 2016

Day	27	9	18	23	15	5	8	12	25	20	2	16	4	6	7	19	11	14
1												x						
2									x	?	?	x	x					
3												x						
5							x											
6							x	?	x		?	x						
7												x						
9				x												x		
10																	x	
12					x											x		
13															x			
14			x															
15	x	x																
19														x				
20			x	?	?	x						x						x
21						x												

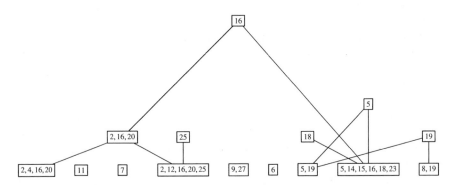

Fig. 2 Diagram of checked points

Due to the number of checking points and the sparse data, we extract the knowledge from Table 4 with the same FCA method used in [8] but reduced the original lattice to show relevant knowledge in Fig. 2. In this Figure we show a diagram of cameras where this vehicle was registered. As we observed previously, location of camera 16 is the most relevant place to analyse criminal reports around the surrounding area, but it turns to camera 5 on middle days of the month.

5 Conclusions and Future Work

In this work we tried to check if there exists in the studied zone similar criminal behaviour patterns that exists in Southern Spain analysing information provided by road traffic cameras. Applying FCA methods we obtain system of implications and concept lattices that give important knowledge about suspects of criminal activities. For this purpose, we detected two examples of vehicles that could be associated to the proposed different patterns.

Our study is not completed due to lack of information, such criminal activities reports and the failure of some cameras that not registered vehicles that pass trough their working areas. For this reason, we can only suggest how to work with the collected data for detecting vehicles that has specified patterns, reducing the size of data that have to be analysed.

With additional information, such professional activity of vehicles, we can discard vehicles with commercial activities and reduce the amount of data that have to be analysed. This reduced data could let us discover systems with implications that related movements of vehicles that follow first pattern applying FCA methods.

Automatic systems that prevent criminal activities could not be developed since we can obtain several suspects according to their movements.

In a near future, all the methods applied in this work will be used in studies about different areas of security, adapting the behaviour patterns to specific scenarios, such airport security, developing new methods that consider unknown values caused by fails in control systems. This study provides a secondary result detecting failures in NPRS analysing different vehicles but not specified what kind of failure exists. Errors in OCR could be improved with the provided information. Also, new similarity relations will be developed that let us check failures in NPRS using fuzzy values in FCA.

Acknowledgements Author want to thank Police forces in Costa del Sol in Southern Spain for their support in this study. This research did not receive any specific grant from funding agencies in the public, commercial, or not-for-profit sectors.

References

1. Florido, E., O. Castaño, A. Troncoso, and F. Martínez-Álvarez. 2015. Data mining for predicting traffic congestion and its application to Spanish data, pp. 341–351. Cham: Springer International Publishing.
2. Ganter, B. 1984. Two basic algorithms in concept analysis. Darmstadt: Technische Hochschule.
3. Guigues, J.L., and V. Duquenne. 1986. Familles minimales d implications informatives resultant d un tableau de donnees binaires. *Mathematiques et Sciences Sociales* 95: 5–18.
4. Lou, Xinyan, Yang Liu, and Xiaohui Yu. 2013. Traffic session identification based on statistical language model, pp. 264–275. Berlin: Springer.
5. Polizia di Stato. 2017. Dataset from traffic cameras.
6. Rodríguez-Jiménez, J.M., P. Cordero, M. Enciso, and A. Mora. 2014. A generalized framework to consider positive and negative attributes in formal concept analysis. In *CLA*, pp. 267–278.

7. Rodríguez-Jiménez, J.M., P. Cordero, M. Enciso, and A. Mora. 2014. Negative attributes and implications in formal concept analysis. *Procedia Computer Science* 31: 758–765.
8. Rodríguez-Jiménez, J.M., P. Cordero, M. Enciso, and A. Mora. 2016. Analyzing criminal networks using formal concept analysis with negative attributes. In *Proceedings of the International Conference on Computational and Mathematical Methods in Science and Engineering, CMMSE.*
9. Rodríguez-Jiménez, J.M., P. Cordero, M. Enciso, and A. Mora. 2016. Data mining algorithms to compute mixed concepts with negative attributes: an application to breast cancer data analysis. *Mathematical Methods in the Applied Sciences* 39: 4829–4845.
10. Rodríguez-Jiménez, J.M., P. Cordero, M. Enciso, and S. Rudolph. 2016. Concept lattices with negative information: A characterization theorem. *Information Sciences* 369: 51–62.
11. R. Wille. 1982. Restructuring lattice theory: an approach based on hierarchies of concepts. In Ordered Sets, ed. I. Rival, pp. 445–470. Boston.
12. Xue, Guangtao, Ke Zhang, Qi He, and Hongzi Zhu. 2012. Real-time urban traffic information estimation with a limited number of surveillance cameras. *Frontiers of Computer Science* 6 (5): 547–559.
13. Zhuang, Peng, Yi Shang, and Bei Hua. 2009. Statistical methods to estimate vehicle count using traffic cameras. *Multidimensional Systems and Signal Processing* 20 (2): 121–133.

Efficient and Accurate Traffic Flow Prediction via Fast Dynamic Tensor Completion

Jinzhi Liao, Xiang Zhao, Jiuyang Tang, Chong Zhang and Mingke He

Abstract Timely and accurate prediction of traffic flow plays an important role in improving living quality of the public, which greatly influences the polices and regulations to be enforced and abided by. In this paper, we focus on urban highway traffic prediction, and present a tensor completion based method, namely, DTC-F. It is conceived on the solid basis of dynamic tensor model for traffic prediction, and in this paper, fast low rank tensor completion and dynamic tensor structure are first combined to pursue high prediction performance. The proposed DTC-F method excavates the inner law of traffic flow data by taking account of multi-mode features, such as daily and weekly periodicity, spatial information, and temporal variations, etc. Empirical evaluation demonstrates the superiority of DTC-F, and indicates that the proposed method is potentially applicable in large and dynamic highway networks.

1 Introduction

Accurate and prompt   prediction of traffic flow information is the key to improving living quality of the urban public. Credible prediction helps traffic department and police arrange transportation planning in order to alleviate traffic congestion and

J. Liao · X. Zhao (✉) · J. Tang · C. Zhang · M. He
National University of Defense Technology, Changsha, China
e-mail: xiangzhao@nudt.edu.cn

J. Liao
e-mail: jzliao@nudt.edu.cn

J. Tang
e-mail: jytang@nudt.edu.cn

C. Zhang
e-mail: czhang@nudt.edu.cn

M. He
e-mail: mkhe@nudt.edu.cn

X. Zhao · J. Tang · C. Zhang · M. He
Collaborative Innovation Center of Geospatial Information Technology, Wuhan, China

© Springer International Publishing AG, part of Springer Nature 2018
F. Leuzzi and S. Ferilli (eds.), *Traffic Mining Applied to Police
Activities*, Advances in Intelligent Systems and Computing 728,
https://doi.org/10.1007/978-3-319-75608-0_6

avoid serious social consequences. For instance, if heavy traffic is foretold to come, traffic wardens may immediately tweak traffic signals, open tidal lane or deploy additional routes. As to traffic police, once heavy traffic flow is predicted, they can arrive at the potential congestion point in advance to maintain traffic order.

In the past, real-time traffic data was difficult to collect, which made data analysis nearly impossible. Nonetheless, with the recent technology booming, proactive traffic management based on data-intensive prediction has been realized. This research also falls into this category and focuses on traffic flow prediction.

Credible traffic flow prediction heavily relies on real-time traffic data. Conventional methods usually model traffic data into time series (1-way) or matrix pattern (2-way) [25]. However, it is observed that traffic data is intrinsically a type of multi-mode information, and existing methods fail to exploit this characteristic. In other words, the major limitation of the 1-way and 2-way methods is that they do not fully leverage the information of traffic data, and only mine patterns in low-dimensional space, losing the developing patterns of the data in high-dimensional space. Intuitively, the more dimensions from which a model handles the data, the more information can be explored, and the better accuracy and reliability the prediction may achieve [24].

Recent effort shows that tensor, the high dimensional expansion of vector and matrix, has an advantage over other structures in explaining multi-mode data [4, 12]. Hence, tensor based method for traffic flow prediction was put forward. With data fixed in corresponding dimension, for example, spatial and temporal modes are expressed in a 2-way tensor, which calculates the similarity in different dimensions, and combines them to get the results. Owing to more dimensions appended, the prediction performance of tensor based methods overmatch the 1-way and 2-way based methods [22]. Lately, many methods originated from tensor decomposition algorithms have been successively applied. In particular, HaLRTC, which was firstly used in image processing, was incorporated for *traffic speed* prediction [18]. Although HaLRTC enjoys good performance in estimating missing data, the method neglects the difference between image processing and traffic prediction, of which the latter contains more behavioral rules. Focusing on *traffic flow* prediction, Tan et al. proposed a dynamic tensor model to predict progressively [24], which is a development of tensor decomposition in completing missing data. Nevertheless, the proposed method is based on Tucker decomposition [11], which fails to completely preserve the patterns of a tensor; that is, each decomposition partially distorts, and hence error accumulates gradually. We address the challenge in this paper by developing a faster and more accurate method that fully leverages multi-mode traffic data to enhance traffic flow prediction.

Inspired by the successful application of tensor in traffic domain, this paper conceives a new dynamic tensor completion method based on fast low rank tensor completion algorithm, namely, DTC-F. Learning from [24], the dynamic tensor model is utilized to perfect our method. In comparison to existing methods, it takes into account more features—spatial mode, week mode, day mode and temporal mode—of traffic data, for better efficiency and accuracy. For empirical evaluation, real-world data is tested by comparing DTC-F with a number of state-of-the-art methods, and the experiment results reveal that **DTC-F** offers remarkable performance gain over the competing alternatives.

Contributions. To summarize, the key contributions of the paper is two-fold:

- We utilize dynamic tensor structure to model traffic data into time series data, and a fast low rank tensor completion algorithm is incorporated to solve the traffic flow prediction problem.
- We experimentally applied the proposed method to prediction problem on real-world traffic flow data, and it is demonstrated to enjoy significant advantage in terms of both efficiency and accuracy.

Organization. After introduction in Sect. 1, Sect. 2 surveys related works on prediction problems in the traffic domain. Necessary background knowledge, and the proposed **DTC-F** method is introduced in Sect. 3. Experimental studies are reported in Sect. 4, followed by conclusion given in Sect. 5.

2   Related Works

With more and more attention drawn into intelligent transportation systems, there have been a large number of methods that were proposed to deal with prediction problems in traffic domain.

The autoregressive integrated moving average (ARIMA) was used to handle the traffic flow prediction problem in early 1970s [1], which then became a widely used technique and was developed persistently. Box–Jenkins time series analyses were applied to predict expressway traffic flow [14]. The ARIMA (0, 1, 1) model was proved to be the most statistically significant for all forecasting. Hamed et al. applied an ARIMA model for traffic volume prediction in urban arterial roads [7]. In succession to these efforts, many variants of ARIMA were devised to improve prediction accuracy, such as Kohonen-ARIMA [26], subset ARIMA [13], ARIMA with explanatory variables [28], vector autoregressive moving average, space–time ARIMA [9], and seasonal ARIMA [29].

Distinct from ARIMA based parametric methods, non-parametric methods were also harnessed in the traffic flow prediction, on account to the stochastic and nonlinear nature of traffic flow. A dynamic multi-interval traffic volume prediction model based on k-NN non-parametric regression was represented in [3]. Functional estimation techniques were applied in a kernel smoother for the auto-regression function to do short-term traffic flow prediction in [6]. A local linear regression model for short-term traffic forecasting was used [20]. Bayesian network was also exploited to forecast traffic flow [21]. Among others, online learning weighted support vector regression was presented for short-term traffic flow predictions [8].

For the weakness of single algorithms, some hybrid methods were explored to enhance prediction performance. For instance, an aggregation approach for traffic flow prediction was proposed based on the moving average, exponential smoothing,

ARIMA, and neural network models [23]. The models were used to obtain three relevant time series that form the basis of a neural network in the aggregation stage. Zargari et al. developed different linear genetic programming, multi-layer perceptron, and fuzzy logic models for estimating 5-min and 30-min traffic flow rates [30]. Cetin and Comert combined ARIMA model with the expectation maximization and cumulative sum algorithms [2]. Recently, an adaptive hybrid fuzzy rule-based system approach was proposed for modeling and predicting urban traffic flow [5].

Beyond all doubts it is difficult for a method to clearly outperform the rest, as different types of predicting problems need distinct models to pinpoint the core, and suitable methods are supposed to handle various realistic demands. To precisely search and explain the latent patterns of the collected data, DTC-F is developed in this research, and proven to be able to obtain promising results.

3   Proposed Method

In this section, we first introduce a dynamic tensor model for traffic flow, and then present a fast tensor completion algorithm for traffic flow prediction.

3.1   Dynamic Tensor Model for Traffic Flow

We first introduce the basics of tensor, and then present the dynamic tensor model for traffic flow.

Tensor basics. Tensor is the extension in mode of vector (1-mode) and matrix (2-mode), which is in essence a multi-way arrays. A n-mode tensor can be defined as $\mathbf{X} \in R^{I_1 \times I_2 \times \dots \times I_n}$, where I_n denotes the quantity of mode-n, and its elements are denoted as $x_{(I_1, \dots, I_k)}$, where $1 \le k \le n$. *Matriculating* operator, which means to unfold a tensor into a matrix, is defined as $unfold\,(\mathbf{X}, n) = X_{(n)}$, in which the tensor element $(I_1, I_2, \dots, I_n)$ is mapped to the matrix element (I_n, J), where

$$J = \prod_{m=1, m \ne n}^{k-1} I_m.$$

Thus, $X_{(n)} \in R^{I_n \times J}$. The reverse of the matriculation is defined as $fold\,(X_{(n)}, n) = \mathbf{X}$ in a similar way.

The inner product of two same-size tensor $\mathbf{A}, \mathbf{B} \in R^{(I_1 \times I_2 \times \dots \times I_N)}$ is defined as the sum of the products of their entries,

$$(\mathbf{A}, \mathbf{B}) = \sum_{i_1} \sum_{i_2} \cdots \sum_{i_N} a_{(i_1, i_2, \dots, i_n)} b_{(i_1, i_2, \dots, i_n)}.$$

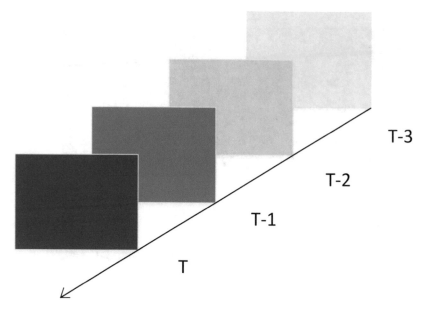

Fig. 1 Tensor stream

For any $1 \leq n \leq N$, the product of a matrix $M \in R^{J \times I_n}$ with a tensor $\mathbf{A} \in R^{(I_1 \times I_2 \times \cdots \times I_N)}$ is expressed as $\mathbf{A} \times_n M$, and transformed into the product of two matrices,

$$\mathbf{Y} = \mathbf{A} \times_n M \Leftrightarrow Y_{(n)} = MA_{(n)}.$$

Denote $\|\mathbf{X}\|_F = \sqrt{(\mathbf{X}, \mathbf{X})}$ as the Frobenius norm of a tensor. It is clear that $\|\mathbf{X}\|_F = \|X_{(k)}\|$.

Tensor stream. Tensor stream can be defined as a series of $(\mathbf{X}_1, \mathbf{X}_2, \ldots, \mathbf{X}_T)$, where each $\mathbf{X}_t \in R^{(I_1 \times I_2 \times \cdots \times I_m)}$, $with 1 \leq t \leq N$. The index of the series is time, as described in Fig. 1. Tensor windows $\mathbf{D}(t, w) = \mathbf{X}_{(T-w+1),\ldots,\mathbf{X}_T}$ with each $\mathbf{X}_t \in R^{I_1 \times I_2 \times \cdots \times I_{m+1}}$ are defined as the combination of fixed number of tensor in the tensor stream.

As traffic flow data possesses temporal-spatial traits, it is intuitive that the adjacent and historical data matters greatly. Therefore, the prediction of traffic flow can be transformed into the completion problem of a dynamic tensor structure.

Dynamic tensor. Recall that our goal is to forecast the traffic flow in the specific time period and the selected location, on the basis of the existing data. As shown in Fig. 2), the time series $(t_i - n\alpha_t, t_i - (n - 1)\alpha_t, \ldots, t_i, \ldots, t_i + m\alpha_t)$ is the time period containing both existing data and missing data, where t_i is the starting time, α_t is the sampling interval, n is the scale of existing data, and m is the predicting horizon of the specific time period. In other words, the historical traffic flow during $(t_i - n\alpha_t, t_i - (n - 1)\alpha_t, \ldots, t_i)$ is used to predict traffic flow of $(t_i + \alpha_t, t_i + 2\alpha_t, \ldots, t_i + m\alpha_t)$. It is observed that such process can be expressed by

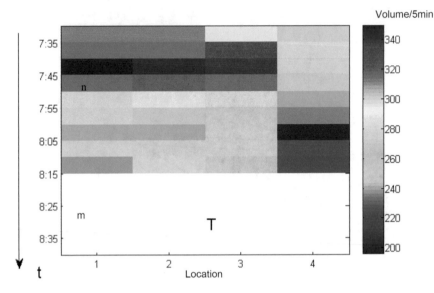

Fig. 2 Volume matrix

$$B^T = f(B^{\overline{T}}),$$

where $T \in R^{l \times m}$, $\overline{T} \in R^{l \times n}$, and $\overline{T}$ is the complement set of T and l represents different locations.

What needs to be appreciated is that the explanatory power of matrix for data is circumscribed. As see in Fig. 2, only the features of sampling intervals and locations are taken into account. The structure express two dimensions of information at most due to the limitation of matrix, and hence, can not seize all pattern of traffic data.

Therefore, we use a 4-way tensor to reconstruct the traffic flow data, which includes sampling intervals, locations, week and days. With the macroscopic variables added, the structure is changed into $\mathbf{B}_t \in R^{l \times w \times d \times s}$, where w is the number of historical weeks, d is the days of historical days (a week has 7 days), and s is the number of intervals $(m + n)$.

Furthermore, dynamic tensor is applied. As mentioned above, n sampling intervals constitute $\mathbf{B}_t^{\overline{T}_t}$, which is used for predicting $\mathbf{B}_t^{T_t}$. The transformation can be expressed as

$$\mathbf{B}_t^{T_t} = f(\mathbf{B}_t^{\overline{T}_t}).$$

In the forecasting phase, the dynamic tensor may update itself by appending the result of $\mathbf{B}_t^{T_t}$ to $\mathbf{B}_t^{\overline{T}_t}$. Thus, the new $\mathbf{B}_t^{\overline{T}_t}$ is defined as $\mathbf{B}_t^{\overline{T}_{t+1}}$, and the size is fixed as same as each tensor windows.

With the missing data predicted, the integrity of existing traffic flow data gets better, and the number and length of the using data gets expanded. For instance, based on history data, the model estimate traffic flow in 7:05. Then with the predicting data put into training set, the model improves its performance in estimating traffic flow in 7:10.

3.2 Fast Dynamic Tensor Completion

We devise an incremental tensor completion algorithm for fast dynamic tensor completion for traffic flow prediction.

Optimization formulation. The essence of common heuristic methods, such as CP decomposition [10] and Tucker decomposition [11], is to transform the high-way tensor into another succinct data structure. Particularly in CP decomposition, a tensor $\mathbf{A} \in R^{n_1 \times n_2 \times \dots n_d}$ is represented with a suitably large r as a linear combination of r rank-1 tensors (vectors); that is,

$$\mathbf{A} = \sum_{i=1}^{r} \lambda_i \alpha_i^1 \otimes \alpha_i^2 \otimes \dots \alpha_i^d.$$

In Tucker decomposition, a tensor $\mathbf{A} \in R^{n_1 \times n_2 \times \dots n_d}$ is decomposed into a set of matrices $U_{(m)} \in R^{I_m \times J_m} (1 \le m \le d)$,and one small core tensor $\mathbf{G} \in R^{J_1 \times J_2 \times \dots \times J_d}$; that is,

$$\mathbf{A} = \mathbf{G} \times_1 U_{(1)} \times_2 U_{(2)} \times \dots \times_d U_{(d)}.$$

But for these two methods, because of transformation of data structure, each decomposition partially distorts, and hence error accumulates gradually.

Different from conventional methods, this paper focuses on incremental tensor completion, which demands fast convergence, high accuracy, and low computational efforts from the algorithm. Thus, we incorporate fast low rank tensor completion (FALRTC) algorithm [15] for this purpose, which has been proven to outperform other tensor completion methods.

Algorithmic solution. For the goal of improving convergence speed and solving the general tensor trance norm minimization problem mentioned above, FALRTC is proposed. When it comes to traffic flow prediction, $\mathbf{D}_t \in R^{J_1 \times J_2 \times J_3 \times J_4}$ for location, week, day and interval mode is the base unit of computing. The main problem is to solve the following optimization,

$$\min_{X_{(t)}} : f(\mathbf{X}) := \sum_{i=1}^{4} \alpha_i \|X_{(i)}\|_*,$$

$$s.t. \ \mathbf{X}_\Omega = \mathbf{D}_\Omega,$$

where α_i's are constants satisfying $\alpha_i \geq 0$, and $\sum_{i=1}^{n} \alpha_i = 1$. The difficulty to efficiently solve the optimization problems lies on the multiple dependent non-smooth terms in the objective function.

Using subgradient information to replace the gradient information can make the convergence rate become $O(K^{-1/2})$ where K is the iteration number [16]. In comparison, the optimal convergence rate for minimizing general smooth functions is $O(K^{-2})$ [16]. Based on a general method proposed to solve a non-smooth optimization problem [17], the basic idea of FALRTC is that (1)convert the original non-smooth traffic flow data into a smooth one; and (2) solve the smooth problem and use its solution to approximate the original problem. Interested readers may refer [15] for more proofs and algorithm details.

In summary, we encapsulate the overall algorithm for completing dynamic tensor structure in Algorithm 1, the performance of which is empirically evaluated in Sect. 4.

Algorithm 1 DTC-F

Require: dynamic tensor $\mathbf{B}_t$, prediction sets $\mathbf{T}_t$, parameters $(\alpha_1, \alpha_2, \ldots, \alpha_n)$;
Ensure: prediction $\mathbf{B}^T$.
  initialize $\mathbf{Z} = \mathbf{W} = \mathbf{B}_t$
  while not converge **do**
    $\mathbf{B}_{t+1} = \text{FALRTC}(\mathbf{B}_t, (\alpha_1, \alpha_2, \ldots, \alpha_n))$
    $t = t + 1$
  end while
  Return $\mathbf{B}^T$;

---

4 Experimental Evaluation

In this section, we report the experimental studies with in-depth analyses.

4.1 Experiment Settings

We used the most widely used public datasets from Caltrans Performance Measurement System (PeMS)[1] for performance evaluation.

Data Sources. Traffic flow data was from PeMS open-access traffic flow datasets, which was collected in real-time from over 39,000 individual detectors. These sensors span the freeway system across all major metropolitan areas of the State of California, USA.

The dataset of south bound freeway SR99, District 10, Stanislaus County, California was utilized in the empirical studies. The index numbers of these detectors

---

[1] http://pems.dot.ca.gov/.

were 1017510, 1017610, 1017710, 1017810, 1017910, 1018110, 1018210, 1018310, 1018410, 1018510, and 1018610 (11 locations in total). The sampling period was from March 1, 2011 to May 29, 2011. The parameters of prediction model were modified on the data from March 1, 2011 to April 15, 2011, and the traffic data from April 16, 2011 to May 30, 2011 were used for evaluating the prediction performance.

Evaluation Indexes. Mean absolute percentage error (MAPE) is used to evaluate the performances of forecasting data. In view of MAPE reducing the error when traffic volumes are higher, the mean absolute error (MAE) is also applied as assessment measures.

$$MAPE = \frac{1}{n} \sum_{t=1}^{n} \frac{|z_t - N_t|}{N_t},$$

$$MAE = \frac{1}{n} \sum_{t=1}^{n} |z_t - N_t|,$$

where z_t is the predicted traffic flow in time t, N_t is the true value of the observation t, and n is the number of predictions.

4.2 Experiment Results

We experimentally compared DTC-F to tensor completion based method with HaL-RTC, Tucker decomposition and CP decomposition. Since the comparing methods (Tucker and CP) are non-convex, multiple, initial points are tested and the average performance is used for comparison. The size of DTC-F is set to 11 (locations)×7 (weeks) ×7 (days)×7 (intervals). For HaLRTC, the value of α_i is set to 1/4 (4 is the mode number). Let $\beta_i = \alpha_i/\gamma_i$, and it has been shown that $\gamma = 100$ owns great performance.

Furthermore, in order to prove the superiority of FALRTC, dynamic tensor structure is applied in the mentioned method to control the disturbance term. In other words, there are 6 different methods put into assessment, HaLRTC, HaLRTC-DT (HaLRTC based on dynamic tensor structure), CP, CP-DT (CP based on dynamic tensor structure), Tucker, Tucker-DT (Tucker based on dynamic tensor structure).

What should not be ignored is that, time scale of prediction models matters. It is unusable for an interval shorter than 3 min, for a decline of predictable information [19]. In the same way, longer intervals will also cause loss of information] [27]. Taking all these into consideration, the aggregation time scale of traffic volume data is set to 5 min in this paper.

The evaluation was conducted in different dimensions, i.e., locations and days, to guarantee the reliability of the experiment results. When one of these dimensions put into optimization procedure, the other parameters remained the same. All experiments were implemented in MATLAB 2013a, and all tests were performed on a PC

Table 1 Accuracy results of locations

Method	MAPE	MAE
DTC-F	**0.037**	**7.989**
HaLRTC	0.403	85.760
HaLRTC-DT	0.180	38.533
CP	0.254	54.064
CP-DT	0.211	49.064
Tucker	0.175	37.322
Tucker-DT	0.112	23.767

Table 2 Accuracy results of days

Method	MAPE	MAE
DTC-F	**0.083**	**14.592**
HaLRTC	0.251	22.211
HaLRTC-DT	0.189	33.120
CP	0.339	57.462
CP-DT	0.133	23.041
Tucker	0.154	28.544
Tucker-DT	0.118	21.564

with Intel Core 2 2.67 GHz and 4GB RAM. Tables 1 and 2 provide the overall results for all the methods involved in the experiments in terms of different dimensions, respectively.

For better appreciation, we also fit and visualize the curves for the locations and days (Monday to Sunday), respectively, in Figs. 3 and 4. It reads from the figures that DTC-F transcends other methods in all dimensions; in other words, DTC-F can help effectively capture the pattern of the traffic data, and hence leads to a better performance.

The performance of CP and Tucker indicates the heuristic algorithm is poor for high rank problems, even when the dynamic tensor structure is applied. In the experiment, when DTC-F and HaLRTC-DT are configured to have the same recovery accuracy, the DTC-F is much faster than HaLRTC-DT, and the disparities in outcome is evident.

Taking both two tables into account, the outcome of same method is obviously discrepant in different modes. The fact indicates that when parameter reference system changed, the utility of tensor completion methods also changed. When compared with original technique, the contrasting method based on dynamic tensor structure own better performance. The observation proves the superiority of dynamic tensor structure. In all conditions, DTC-F gets the most promising result.

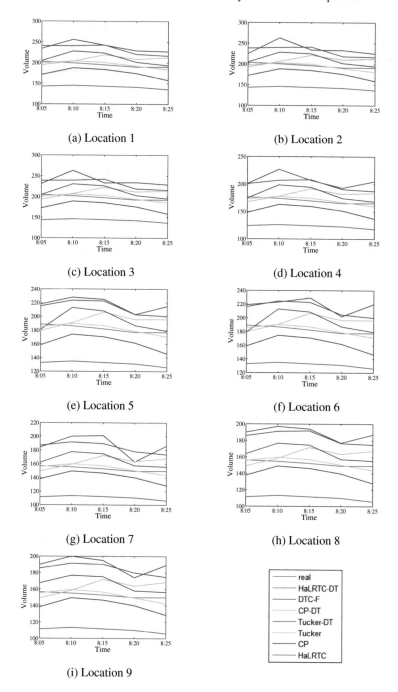

Fig. 3 Fitted curves for locations

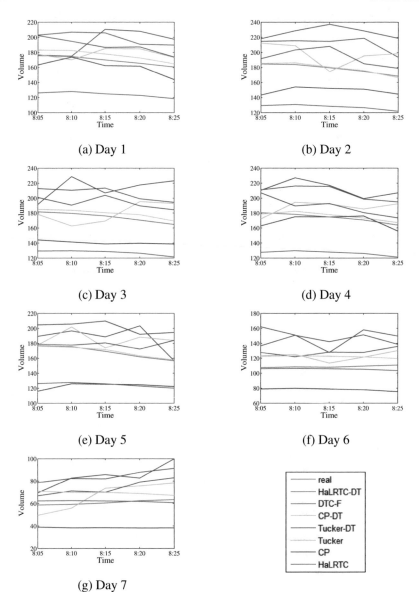

(a) Day 1

(b) Day 2

(c) Day 3

(d) Day 4

(e) Day 5

(f) Day 6

(g) Day 7

Fig. 4 Fitted curves for locations

5 Conclusion

In this paper, a tensor completion method DTC-F is proposed to solve the flow prediction problem in traffic domain. Fast low rank tensor completion (FALRTC) and dynamic tensor structure are first combined to achieve the high-performance of prediction model. DTC-F can excavate the inner law of traffic flow data, since multi-mode features are taken into model. Multi-view experiments prove that DTC-F's superiority, and imply the application prospect in large and dynamic traffic environment. During the research, it has been brought to our attention that the model gets superior outcome in 5 minutes interval. Nevertheless, when it comes to long-term traffic flow prediction, the effect of DTC-F is yet to be evaluated. As future work, we plan to further experiment with our proposal; currently, the experiments were carried out on PeMS data, and we plan to apply DTC-F to TRAP-2017 benchmark dataset to see its advantages and limitations.

Acknowledgements This work was in part supported by NSFC under grants Nos. 61402498, 71331008 and 71690233.

References

1. Ahmed, M.S., and A.R. Cook. 1979. Analysis of freeway traffic time-series data by using box-jenkins techniques. *Transportation Research Record* 722: 1–9.
2. Cetin, M., and G. Comert. 1965. Short-term traffic flow prediction with regime switching models. *Transportation Research Record* 23–31: 2006.
3. Chang, H., Y. Lee, B. Yoon, and S. Baek. 2012. Dynamic near-term traffic flow prediction: System oriented approach based on past experiences. *IET Intelligent Transport Systems* 6 (3): 292–305.
4. Cichocki, Andrzej, Rafal Zdunek, Anh Huy Phan, and Shun-ichi Amari. 2009. *Nonnegative Matrix and Tensor Factorizations—Applications to Exploratory Multi-way Data Analysis and Blind Source Separation*. Wiley.
5. Dimitriou, Loukas, Theodore Tsekeris, and Antony Stathopoulos. 2008. Adaptive hybrid fuzzy rule-based system approach for modeling and predicting urban traffic flow. *Transportation Research Part C: Emerging Technologies* 16 (5): 554–573.
6. El Faouzi, N.E. 1996. Nonparametric traffic flow prediction using kernel estimator. In *Proceedings of The 13th International Symposium on Transportation and Traffic Theory*, 41–54.
7. Hamed, M., H. Al-Masaeid, and Z. Said. 1995. Short-term prediction of traffic volume in urban arterials. *Journal of Transportation Engineering* 121 (3): 249–254.
8. Jeong, Young-Seon, Young-Ji Byon, Manoel Mendonca Castro-Neto, and Said M. Easa. 2013. Supervised weighting-online learning algorithm for short-term traffic flow prediction. *IEEE Transactions of Intelligent Transportation Systems* 14(4): 1700–1707.
9. Kamarianakis, Y., and P. Prastacos. 1857. Forecasting traffic flow conditions in an urban network: Comparison of multivariate and univariate approaches. *Transportation Research Record* 74–84: 2003.
10. Kolda, T.G., E. Acar, D. M. Dunlavy, and M. Mørup. 2011. Scalable tensor factorizations for incomplete data. *Chemometrics Intelligent Laboratory Systems* 106(1): 41 – 56.
11. Kolda, T., and B. Bader. 2009. Tensor decompositions and applications. *SIAM Review* 51 (3): 455–500.

12. Kolda, Tamara G., and Brett W. Bader. 2009. Tensor decompositions and applications. *SIAM Review* 51 (3): 455–500.
13. Lee, S., and D. Fambro. 1999. Application of subset autoregressive integrated moving average model for short-term freeway traffic volume forecasting. *Transportation Research Record* 1678: 179–188.
14. Levin, M., and Y.-D. Tsao. 1980. On forecasting freeway occupancies and volumes. *Transportation Research Record* 722: 47–49.
15. Liu, Ji, Przemyslaw Musialski, Peter Wonka, and Jieping Ye. 2013. Tensor completion for estimating missing values in visual data. *IEEE Transactions on Pattern Analysis and Machine Intelligence* 35 (1): 208–220.
16. Nesterov, Y. 1998. *Introductory lectures on convex programming*, 119–120. Notes: Lecture.
17. Nesterov, Y. 2005. Smooth minimization of non-smooth functions. *Mathemtaical Programming* 103 (1): 127–152.
18. Ran, Bin, Huachun Tan, Jianshuai Feng, Ying Liu, and Wuhong Wang. 2015. Traffic speed data imputation method based on tensor completion. *Computational Intelligence and Neuroscience* 364089:1–364089:9.
19. Ritchie, S.G., and C. Oh. 2005. Exploring the relationship between data aggregation and predictability to provide better predictive traffic information. *Transportation Research Record* 1935 (1): 28–36.
20. Sun, H.Y., H.X. Liu, H. Xiao, R.R. He, and B. Ran. 1836. Use of local linear regression model for short-term traffic forecasting. *Transportation Research Record* 143–150: 2003.
21. Sun, Shiliang, Changshui Zhang, and Yu. Guoqiang. 2006. A bayesian network approach to traffic flow forecasting. *IEEE Transactions on Intelligent Transportation Systems* 7 (1): 124–132.
22. Tan, Huachun, Guangdong Feng, Jianshuai Feng, Wuhong Wang, Yu-Jin Zhang, and Feng Li. 2013. A tensor-based method for missing traffic data completion. *Transportation Research Part C: Emerging Technologies* 28:15 – 27. Euro Transportation: selected paper from the EWGT Meeting, Padova, September 2009.
23. Tan, Man-Chun, Sze Chun Wong, Jian-Min Xu, Zhan-Rong Guan, and Peng Zhang. An aggregation approach to short-term traffic flow prediction. *IEEE Transactions on Intelligent Transportation Systems*, 10(1):60–69.
24. Tan, Huachun, Wu Yuankai, Bin Shen, Peter J. Jin, and Bin Ran. 2016. Short-term traffic prediction based on dynamic tensor completion. *IEEE Transactions on Intelligent Transportation Systems* 17 (8): 2123–2133.
25. Tchrakian, Tigran T., Biswajit Basu, and Margaret O'Mahony. 2012. Real-time traffic flow forecasting using spectral analysis. *IEEE Transactions on Intelligent Transportation Systems* 13 (2): 519–526.
26. Van Der Voort, Mascha, Mark Dougherty, and Susan Watson. 1996. Combining kohonen maps with arima time series models to forecast traffic flow. *Transportation Research Part C: Emerging Technologies* 4 (5): 307–318.
27. Vlahogianni, E., and M. Karlaftis. 2011. Temporal aggregation in traffic data: Implications for statistical characteristics and model choice. *Transportation Letters* 3 (1): 37–49.
28. Williams, B.M. 2001. Multivariate vehicular traffic flow prediction: Evaluation of arimax modeling. *Transportation Research Record* 1776: 194–200.
29. Williams, B.M., and L.A. Hoel. 2003. Modeling and forecasting vehicular traffic flow as a seasonal arima process: Theoretical basis and empirical results. *Journal of Transportation Engineering* 129 (6): 664–672.
30. Zargari, Shahriar Afandizadeh, Salar Zabihi Siabil, Amir Hossein Alavi, and Amir Hossein Gandomi. 2012. A computational intelligence-based approach for short-term traffic flow prediction. *Expert Systems* 29(2): 124–142.

Reducing the Risk of Accidents with Not Insured British Vehicles in Southern Spain

Jose Manuel Rodriguez-Jimenez, Jesus Cabrerizo, Dario Perez and Ignacio Sanchez

Abstract This research commenced when different Police departments from the Southern Spain observed an increase of irregularities in accident reports involving British vehicles. Primarily, vehicles seemed to have valid insurance, however, since they lacked Administrated Licences (Road Tax) or approved Roadworthy Certification, the damage caused in the accident by these vehicles was not being covered by Insurance Companies. British Police offers public information about British vehicles via websites for government and companies that Spanish Police checks with possible fraudulent vehicles to avoid this situation. This fraud affects countries and insurance companies. There are detected a significant amount of fraudulent vehicles that are driven in Spain. Using the information provided by British government this number of fraudulent vehicles and the problem related to accidents with non insured vehicles has been reduced. Public information and collaboration between British and Spanish Polices allow detecting frauds in vehicles with administrative irregularities and protect drivers that have accidents with these vehicles ensuring safety for road users. Methodology used in this research could be extended to different European Police departments, not only Spanish, where fraudulent vehicles are detected. This also will constitute a global reduction in risk of accidents due to mechanical problems in vehicles that are not officially checked.

J. M. Rodriguez-Jimenez (✉)
Andalucia Tech, University of Malaga, Malaga, Spain
e-mail: jmrodriguez@ctima.uma.es

J. M. Rodriguez-Jimenez
Mijas Police Department, Malaga, Spain

J. Cabrerizo
Tarifa Police Department, Cádiz, Spain

D. Perez
San Roque Police Department, Cádiz, Spain

I. Sanchez
Seville Police Department, Seville, Spain

© Springer International Publishing AG, part of Springer Nature 2018
F. Leuzzi and S. Ferilli (eds.), *Traffic Mining Applied to Police Activities*, Advances in Intelligent Systems and Computing 728,
https://doi.org/10.1007/978-3-319-75608-0_7

1 Introduction

When foreign residents move into a new community, they carry all their belongings, in particular, vehicles. After living six months in Spain, they must replace the original registration number of their vehicles by a Spanish plate. The change procedure for the Spanish registration number is expensive and sometimes owners do not change it because they think that this irregularity will never be detected by Spanish Police who is assumed to be unable to obtain detailed information about foreign vehicles. A high percentage of resident citizens deny that they have their home in Spain because, in this case, they must replace the registration number.

Other vehicles belong to partial residents (i.e. people who have business in different countries, live in Southern Europe just in winter, etc.) who park old vehicles in the secondary residence in Spain.

Since the European Union eliminated Border Checks for people and vehicles, hence became easier to travel around Europe, this situation meant new benefits for European Citizens, but also made it easier for the movement of the import/export of illegal or stolen vehicles between the European countries without border control [2, 10, 16].

The structure of Vehicle databases differs for each country, the fact that these databases are not connected become an obstacle which the Police forces discover in their investigations of detecting this type of illegal activities.

Some criminal organizations are hidden behind these import/export activities covered by the uncertainty plus the presumption that the driver of a foreign vehicle is only a tourist [3].

Public forces that protect The Law are allowed to access the data for National Citizens for the purpose to ensure safety and security of the community. But they cannot access data about foreign people or vehicles by the same means. Regarding ownership of a driver licence, due to some recent treaties, for example Prüm Treaty [14], this problem has been reduced but not solved. Several discrepancies have been found by Police forces for some people between their National Database from their home country (where they usually live) and Spanish Database System.

One of the most common and important problems is that the documentation is not standard in Europe and wrote in the homeland language, so it is difficult to decide about their legality. By accessing foreign vehicle databases, Police can check the vehicle status and detect irregularities that constitute fraud. However, not all information is available even for the Police forces, because some countries prefer not to disclose personal details. Other countries also protect confidential information about vehicle ownership (name and address), information that could be very useful to avoid fraud, but do provide other facts whether the vehicle is licensed, insured roadworthy or stolen. With appropriated handling, we can reduce it [4, 5].

As we mentioned in the Abstract, this research began when different Police departments from the Southern Spain observed an increase of irregularities in accident reports involving British vehicles. Primarily, vehicles seemed to have valid insurance, however, since they lacked Administrated Licences (Road Tax) or approved

Roadworthy Certification, the damage caused in the accident by these vehicles was not being covered by Insurance Companies.

The British law prevents pay road tax if you do not have a valid British Roadworthy Certification (Spanish certifications are not valid), so a vehicle that stays long time in Spain loses its authorization to circulate.

In 2009, European Union specifies in its Parliament Directive [13] the conditions related to insurance against civil liability with respect to the use of motor vehicles, and the enforcement of the obligation to insure against such liability. This directive affects foreign vehicles that are driven in the European Union and cover themselves in case of an accident. In one of its article, it specifies that the responsibility for unregistered vehicles is carried out by the country in which the vehicle is usually located. It means that, if a vehicle that causes an accident is not properly registered in its original country, even though it has a registration number from this country, the damage has to be supported by the country where the accident took place.

It is also specified that it is prohibited to do specific or discriminatory checks about insurance from others countries, but it allows that "they may carry out non-systematic checks on insurance provided that those checks are not discriminatory and are carried out as part of a control which is not aimed exclusively at insurance verification".

This situation constitutes a critical problem because there are some areas where numerous foreign not licensed vehicles are located and countries cannot control them effectively. The increasing number of accident reports where a British unlicensed vehicle is involved lead us to focus on discovering if there exists a real problem of fraud on exported vehicles, because sometimes, at a first glance, the check of documents indicates only apparently a valid insurance.

The case of Road Worthy Test (RWT) is different and there are European Union directives [6, 7] that unify the criteria for these control tests. It is assumed, in the proposed actions of these directives, the recommendation of checking vehicle mechanical status. It is recommended by the European Union to check that vehicles have a valid RWT regardless of the country where the vehicles are located in.

These different laws lead to a situation where Police forces can report a vehicle with no valid RWT, but they cannot do the same with a vehicle that has no valid insurance. This situation represents a relevant trouble for both countries, where the vehicle is licensed and where it is exported.

In the absence of an adequate bibliography about this issue, we contact policemen from different countries who are experts in fake documents. They assist us reporting the existence of public websites, that constitute an alternative to official "EUropean CAR and driving licence Information System" (EUCARIS) that is not public and where British vehicles are not included. These websites are useful tools for the law enforcement community in that they help to prevent fraud and to fight against criminal activities related to the free flow of vehicles in the European Union.

There are studies about measuring fraud in a whole way [9, 12]. We mix data from administrative records and sample surveys with the purposes of policy making and generate strategic solutions to avoid the fraud and define what kind of fraud we want to investigate, focusing this research on administrative fraud in vehicles by evading payment of taxes related to fraud in insurance or vehicle inspections, in which a vehicle is inspected to ensure that it conforms to regulations governing safety, emissions, or both. We face the problem that if the fraudulent licence-holder maintains a perfect driving record, then the visible marginal cost of the fraud is not great, but it exists, and we need to detect it because it could be a potential safety problem for other drivers and vehicles.

EUCARIS stands for EUropean CAR and driving licence Information System. It is a unique system, developed by and for governmental authorities, that provides opportunities for countries to share their vehicle and driving licence registration information and/or other transport related data helping to fight against vehicle theft and registration fraud. EUCARIS is not a database but an exchange mechanism that connects the Vehicle and Driving Licence Registration Authorities in Europe, but Great Britain is not connected so we cannot use it in our research.

In Sect. 2, we describe the methodology used detailing how we collect the sample and the weakness of this methodology. Section 3 shows the results and the discussion about the data obtained in the city of Mijas. Some practical applications are mentioned in Sect. 4. We finish this work with the conclusions and future work of this research.

2 Methodology

There are six fields that we remark as important information:

- Maker (manufacturer of the vehicle)
- Vehicle Identification Number or chassis number (VIN)
- Licence (if the vehicle has a valid licence to be driven)
- Roadworthiness (if the vehicle has a positive technical control)
- Insurance (if the vehicle is insured)
- Stolen (if the vehicle is registered as stolen)

Maker, VIN and stolen are fields that prevent fraud with stolen vehicles. Fraud in documents is related to no-licensed vehicles (untaxed, declared off-road or exported vehicle), expired RWT (Road Worthiness Test) or vehicles without a valid insurance.

Since we contacted British Police, they give us an important set of websites where Police forces can find the information required to avoid the use of fraudulent vehicles. Several public data about vehicles can be checked just knowing the registration number. These websites can be accessed freely via smartphones or laptops so there are easy tools to work with them. These websites are detailed in Table 1.[1]

The VIN is a data that websites usually doesn't show due to security purposes because thefts can modify documents with valid VINs that are not registered as stolen. We cannot use it in our research.

---

[1]DVLA: https://www.gov.uk/government/organisations/driver-and-vehicle-licensing-agency;
MID: https://ownvehicle.askmid.com/; TotalCarCheck: https://totalcarcheck.co.uk/; UKVehicle:
https://www.gov.uk/vehicle-tax.

2.1   Collecting the Samples

Police officers avoid disturbing drivers with British vehicles involved in this research unless the suspect of a fraudulent vehicle is detected. This is one of the main proposals for the researchers due to ethical reasons because this is not a chase on British vehicle drivers, but try to solve a safety problem.

The research was developed in 4 phases:

1. Police collected a list of registration numbers of British vehicles to check status of vehicles and their locations in different Police reports from accidents, fined vehicles, or security control points. There are stored in a database the fields corresponding with registration number, date and location for each vehicle. Due to legal reasons, no data of the owner or driver is stored.
2. The administrative statuses of these vehicles were checked with the help of the information provided by the different websites that have been previously described in Table 1. The initial analysis of the data provided in this list of vehicles, allowed Police to do an initial assessment about possible fraudulent vehicles and where they were located.
3. Police carried out exhaustive controls on British vehicles with anomalies (not all British vehicles) that appear in the previous list to compare the information provided by the websites and the documents of the vehicle. Controls were made in different areas without repeating the same control points in consecutive days. The aim of these controls was to ensure that the information provided by the websites was correct.
4. Final statistical analysis of the collected data for detecting if there exists fraud and the types of fraud committed was developed. With this information could be verified if there exists a real safety problem related to non insured vehicles due to administrative anomalies. It is said that a vehicle has an anomaly if it is not properly licensed, has an invalid RWT or it is not insured. If the problem with the vehicle is solved (only the payment of the road tax because RWT cannot be solved in Spain) or it is moved off the road by the Police, it is said that the problem is solved. Also, if it is detected that the system does not give an answer to a query it is said that the system fails.

Table 1 Information provided by websites about British vehicles

Website	Maker	VIN	Licence	RWT	Insurance	Stolen
DVLA	X		X	X		
MID	X				X	
TotalCarCheck	X		X	X		X
UKVehicle	X					

2.2 *Weakness*

There are two facts that we have to consider in this methodology.

The first one is that Police officers can store registration numbers in different places, but it is so difficult to assure that the same vehicle can be register again to check the validity of the data. Due to different reasons, some vehicles could be never being registered in the control points, so there is a hidden fraud that we have to consider.

The second one is that registers of vehicles are dynamic data and can change their values in a few days so stored data could be not correct. There are stored different values for the same registration number so we only take into account the last one. This is only important due to statistical purposes because when a vehicle is checked in a security control point the information used is the actual one.

3 Results and Discussion

Different cities in Southern Spain began this research at the same time but not all of them collected a representative sample to do a proper research. The initial percentages of vehicles with administrative anomalies are quite similar, but we look for a representative sample to do a proper research.

There is a relevant difference between residents and partial residents citizens. Most of the vehicles in the sample belong to the second category, old vehicles that stay in Spain and are only used on holidays.

3.1 *A Real Case of Study in Mijas (Spain)*

This city was selected because of the high percentage of accident involving British vehicles over the total number of British people living there. Police officers involved in this research have a solid experience and knowledge about foreign vehicle plates and documents.

The research began on 1-1-2015 and finished on 31-7-2016, however, Mijas Police Department is still working on this problem to reduce it.

In the first month (January 2015) we only collect information from reports of accidents and fined vehicles. This initial sample was small but very representative, so we decided to increase the sample size with other vehicles detected in random check points.

In the first six months, the initial sample was collected and the information from websites was used in security control points. Table 2 resume the data collected. It has been made with the aim to prove the efficiency of the information from the websites previously described as complementary tools for Police forces. During this period, an

Table 2 Sample 1. Data from 1-1-2015 to 30-6-2015

Month	Vehicles	Unlicensed	No RWT	Not insured	Solved	Fail
Jan 2015	13	11	10	6	3	0
Feb 2015	211	91	66	22	24	1
Mar 2015	161	66	48	26	13	2
Apr 2015	325	133	109	48	24	2
May 2015	259	81	61	30	18	5
Jun 2015	94	30	25	2	2	1
Total*	807	323	246	108	84	9
%		40.02	30.48	13.38	10.41	1.11

*Vehicles that appear twice or more are reduced to one register

initial sample of 807 vehicles was stored. All the information provided by websites was added to the database.

Despite having administrative information from websites, Police forces never act on driverless vehicles parked on the public roads. This information is guaranteed by the British government, but it is always safer to confirm with documents. For this reason, some fraudulent vehicles are detected, but their anomalies remain not solved.

Mijas Police Department detects only 1.11% of failures in the websites checked. Most of them are related with older vehicles that are not properly registered in their national databases or maker is not properly stored in the British database (for example, website suggest Maker "Ford" and there are cases where you have to input "Ford king ranch") and don't give results to users. When websites do not offer an answer to queries, possible problems are studied and possible mistakes writing the number of plate in a control are discarded using Levenshtein's distance [11] with the registration numbers stored. So, in this way, human errors are dismissed and the possibility of a false plate is considered. With this amount of failures and the mentioned reasons we accept that the information given by websites is trustworthy.

The percentage of unlicensed vehicles and without valid RWT, 40.02 and 30.48% respectively, were so high so we can assert that there exists a real problem related to fraudulent vehicles. Not insured vehicles represent 13.38% from the sample, that is also another problem that Police forces cannot prosecute due to European laws.

Results from this sample were shown to insurance companies (the percentage of unlicensed vehicles, without valid RWT and not insured). They were asked about this situation, and they verify that, for them, the cost was expensive for their clients and for the companies that have to increase the price of the different policies to cover this kind of accidents in a national guarantee fund. Their data show that the main cause of these accidents are mechanical failures that are related to vehicles without a valid RWT, most of them old cars. However, not insured vehicles are covered by their national governments and, for insurance companies, there are no problems related to these vehicles, only with vehicles unlicensed and without valid RWT. For this reason, in the following sample, we discard to check and store if the vehicle has a registered insurance.

Table 3 Sample 2. Data from 1-7-2015 to 31-7-2016

Month	Vehicles	Unlicensed	No RWT	Solved	Fail
Jul 2015	167	58	34	14	2
Aug 2015	114	37	27	13	2
Sep 2015	196	61	46	25	1
Oct 2015	78	32	23	12	1
Nov 2015	194	57	37	18	1
Dec 2015	210	69	53	27	0
Jan 2016	229	54	38	22	1
Feb 2016	260	61	41	27	3
Mar 2016	233	43	30	19	5
Apr 2016	260	42	24	14	3
May 2016	235	45	45	18	0
Jun 2016	180	32	25	13	2
Jul 2016	145	36	23	14	0
Total*	1606	455	333	235	20
%		28.33	20.73	14.63	1.25

*Vehicles that appear twice or more are reduced to one register

Once it is checked that information provided by websites is trustworthy and there is an extended real problem with fraudulent vehicles, Mijas Police Department uses this information to reduce the number of these vehicles that are driven in public roads because they are considered a problem for the safety of road users.

Data from 1-7-2015 to 31-7-2016 was stored in the same way to analyse the evolution of the number of fraudulent vehicles and evaluate the work of Police forces. 4 police officers collected data in security control points located in 9 previously designed places (commercial areas and residential areas). Table 3 shows these data.

An analysis of data stored in the samples of Mijas Police Department has to be developed. There exists data from other Spanish cities, but the size of the sample is smaller (less than 100 vehicles) and not comparable, so researchers decided to discard these comparisons.

Comparing the initial percentage of unlicensed vehicles from the first sample, 40.02%, with the same percentage in the second sample, 28.33%, there exists and effective reduction of these fraudulent vehicles. Also, if we discount "solved" vehicles, this percentage is 13.70%, that shows a 65.77% reduction in the number of cases located.

Figure 1 shows the percentage of unlicensed vehicles by months. There can be observed a descending estimated linear regression line that confirms how the number of unlicensed vehicles are decreasing. Also, can be observed that there is a low increase in the percentage of unlicensed vehicles from October 2015 to December 2015 because in this season of the year British partial residents come to Spain on holidays and "resident" vehicles are driven on public roads.

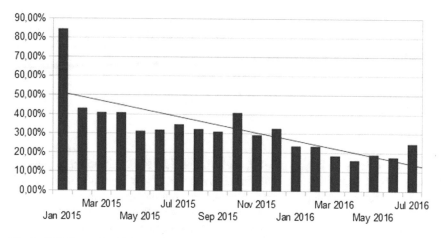

Fig. 1 % Unlicensed vehicles by months with estimated linear regression line

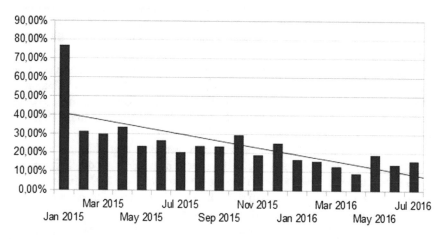

Fig. 2 % Vehicles without RWT by months with estimated linear regression line

In the case of not valid RWT, the initial percentage of 30.48% was reduced to 20.73% if we compare the first sample with the second one. The real percentage is less than this one, but this kind of administrative fault have to be solved in Great Britain, and we cannot obtain this data. This problem is related with unlicensed vehicles (a vehicle with not valid RWT cannot be licensed) but an approximate percentage cannot be provided. Since the mechanical status of these vehicles is not controlled properly, this reduction could be associated with a reduction in the risk of accidents.

Figure 2 shows the percentage of vehicles without RWT by months. In the same way as happens in Fig. 1, there can be observed a descending estimated linear regression line that confirms how the number of vehicles without RWT are decreasing. The percentage of vehicles without RWT is always smaller than unlicensed vehicles.

Table 4 Sample 2. Data from 1-1-2015 to 31-7-2016

Country	Vehicles	Unlic.	% Unlic.	No RWT	% No RWT	Solved
Finland	104	17	16.35	24	23.08	2
The Netherlands	251	17	6.77	32	12.75	16
Sweden	203	22	10.84	73	35.96	11

These vehicles without control of their mechanical status constitute a potential risk due to possible fails that could cause an accident, so it is an important benefit that Police forces can detect them and move them out on public roads.

Since was reported a safety issue because there exists British vehicles with administrative anomalies there was a research about analogous reports with vehicles from other countries, but these cases are not detected. Contacting with foreign Police forces, we obtain websites from other 3 countries: Finland [1, 8], The Netherlands [15] and Sweden [17]. We resume the data collected from these countries in Table 4.

The sizes of these samples are smaller than British samples so it is difficult to compare results efficiently.

In this sample we can check that the problem with fraudulent vehicles also exists in other countries, but there are no reports due to the small amount of fraudulent vehicles so the risk of accidents is low, but exists. It is significant that in these cases there are more vehicles without RWT than unlicensed because in these countries RWT and road tax are not related.

In the case of Netherlands, a road tax to keep the vehicle licensed it is not necessary, so all the registered unlicensed vehicles are exported vehicles that don't change the registration number to Spanish one. They are easy to detect due to the different plate (white, not orange) but some of them try to do false copies emulating original plates. This is the reason for the big percentage of solved cases.

We asked randomly 50 drivers of fraudulent vehicles about the reason of the status of the vehicle and the answers were:

- 36% Don't know the reason or prefer not to answer.
- 20% Don't have enough money to change to Spanish registration number.
- 18% Don't want to change her/his old vehicle to a left-hand-drive vehicle.
- 12% Both previous answers.
- 10% Don't know that have to pay the British road tax in Spain.
- 4% Wants to save money and doesn't think that Spanish Police will detect the fraud.

It is not possible to confirm any hypothesis about why this fraud exists. The highest percentage of drivers prefer not to answer that question, it seems that for some of them it is a shameful situation because they are not criminals.

Table 5 Sample 3. Data from 1-7-2017

Date	Vehicles	Unlicensed	No RWT	Solved	Fail
1 Jul 2017	31	3	2	2	0
%		9.68	6.45	6.45	0.0

3.2 Actual Results

Actually, we can observe that the number of British old vehicles are reduced and there exists a growth in the number of new vehicles (5 years old or newer). This situation reduces the number of unlicensed vehicles and vehicles with not valid RWT.

We collect a sample to check actual data about British vehicles on the first day of July 2017 and the results are displayed in Table 5. Registered unlicensed vehicles are under 10% (9.68%) and vehicles with not valid RWT also were reduced to 6.45% of registered vehicles in the same period, so both problems are efficiently reduced.

3.3 Other Results

The information provided by websites allows Police forces not only to detect fraudulent vehicles, but to report 2 stolen vehicles, 2 vehicles with false plates and 9 false documents issued technical inspections while the vehicles were seized.

4 Practical Applications

The methodology proposed in this research has proven to be efficient in reducing a given real problem of safety so it could be extended to different Police departments, not only in Spain but in other countries.

Different cities in Southern Spain began their own research with different results due to the amount of British vehicles registered but all of them achieved a reduction in the number of fraudulent vehicles, specially old vehicles without RWT, increasing safety in public roads. This will constitute a global reduction in risk of accidents due to mechanical problems in vehicles that are not officially controlled by their governments.

In the case of Southern Spain, there are a significant number of British citizens that live in this zone. There are other places where the largest colony of foreign citizens could be German, French, etc. and other public websites are needed.

5 Conclusions and Future Work

Websites with free access to vehicle data from other countries do it possible to achieve important data in a fast and secure way. These websites help Police to avoid fraud not only in Spain, but in the European Union. The high percentage of fraud detected in the real case study in the city of Mijas shows that there exists a real problem. Police can fight against fraudsters if they have the necessary tools. Focus all Police efforts in prosecuting foreign vehicles is not inferred from the results of this research despite the detected fraud, because it could be a generalization and most of them are not fraudsters or criminals. Maybe we do not need better databases for obtaining information about foreign vehicles, but an easy way to access them could be useful.

In our comparative evolution of fraud since this research started, we show that these websites have an important relevance into policing. Administrative fraud reduction is possible easily ensuring the safety of potential victims. The risk of accidents has been reduced since vehicles that are driven without any kind of control of their mechanical status are detected.

The benefits of this public information are that the estimated cost of prosecuting fraud is not expensive using public tools. We can protect victims of these frauds easily. Potential victims are not only located in the country where the vehicles are moving, but also the original countries where different taxes are not paid have an indirect cost related to possible accidents.

Protection of personal data from its national law is one of the handicaps that Police officers found in the prosecution of fraud. This research has shown how the Police can face fraud in a more agile and effective way if governments offer tools that let Police to access specific data.

As a representative example of free access online tools with relevant information about vehicles, we remark website from Great Britain Government that allows their own citizens to fight against fraud, collaborating with reports that locate and discover fraudsters.

In a future work, we will continue this research in foreign vehicles from other countries to avoid fraudulent traffic of exported vehicles with administrative faults.

There was a project that uses Automatic Number Plate Recognition (ANPR) and classify registration numbers by countries, connecting with the respective websites to extract information about their legal status, using tablets. This project was actually suspended.

Acknowledgements Authors want to thank Police forces in Costa del Sol in Southern Spain for their support in this research. Special thanks to Dr. Inmaculada de las Peñas, from Applied Mathematics Department at University of Malaga, for her special support in the revision of this paper and her suggestions. This research did not receive any specific grant from funding agencies in the public, commercial, or no profit sectors.

References

1. A-katsastus. A-Katsastus.fi, n.d.
2. Boswell, C., and A. Geddes (eds.). 2011. *Migration and Mobility in the European Union.* Houndmills, Basingstoke, Hampshire: Palgrave Macmillan.
3. Clarke, R.V., and R. Brown. 2003. International trafficking in stolen vehicles. *Crime and Justice* 30: 197–227.

4. Dionne, G., and C. Laberge-Nadeau (eds.). 1999. *Automobile Insurance: Road Safety, New Drivers, Risks, Insurance Fraud and Regulation*. Boston: Kluwer Academic.
5. Dionne, G., and K.C. Wang. 2013. Does insurance fraud in automobile theft insurance fluctuate with the business cycle? *Journal of Risk and Uncertainty* 47: 1.
6. European Parliament. Directive 2009/40/EC on roadworthiness tests for motor vehicles and their trailers. 2009.
7. European Parliament. Directive 2014/45/EU on periodic roadworthiness tests for motor vehicles and their trailers and repealing Directive 2009/40/EC Text with EEA relevance. 2014.
8. Finish Transport Safety Agency. Trafi Liikenteen turvallisuusvirasto, n.d.
9. Gee, J., and M. Button. 2015. *The Financial Cost of Fraud*. London: PKF Littlejohn.
10. Gerber, J., and M. Killias. 2003. The transnationalization of historically local crime: Auto theft in western europe and russia markets. *European Journal of Crime, Criminal Law and Criminal Justice* 11 (2): 215–226.
11. Levenshtein, V.I. 1966. Binary codes capable of correcting deletions. *Insertions and Reversals. Soviet Physics Doklady* 10: 707.
12. Levi, M., and J. Burrows. 2008. Measuring the impact of fraud in the uk: A conceptual and empirical journey. *British Journal of Criminology* 48 (3): 293–318.
13. Merkin, R., and J. Steele. 2013. *Insurance and the Law of Obligations*. Oxford: Oxford University Press.
14. Miettinen, S. 2013. *Criminal Law and Policy in the European Union*. Abingdon, Oxon: Routledge Research in EU Law.
15. Netherlands Traffic Authority RDW. RDW, n.d.
16. Reichel, P., and J. Albanese. 2005. *Handbook of Transnational Crime and Justice*. Thousand Oaks, California: Sage Publications.
17. Swedish Transport Agency. Fordonsuppgifter, n.d.

Unsupervised Classification of Routes and Plates from the *Trap-2017* Dataset

Massimo Bernaschi, Alessandro Celestini, Stefano Guarino,
Flavio Lombardi and Enrico Mastrostefano

Abstract This paper describes the efforts, pitfalls, and successes of applying unsupervised classification techniques to analyze the Trap-2017 dataset. Guided by the informative perspective on the nature of the dataset obtained through a set of specifically-written perl/bash scripts, we devised an automated clustering tool implemented in python upon openly-available scientific libraries. By applying our tool on the original raw data it is possibile to infer a set of trending behaviors for vehicles travelling over a route, yielding an instrument to classify both routes and plates. Our results show that addressing the main goal of the Trap-2017 initiative (*"to identify itineraries that could imply a criminal intent"*) is feasible even in the presence of an unlabelled and noisy dataset, provided that the unique characteristics of the problem are carefully considered. Albeit several optimizations for the tool are still under investigation, we believe that it may already pave the way to further research on the extraction of high-level travelling behaviors from gates transit records.

1 Introduction

The advances and widespread availability of automatic Number Plate Reading Systems (NPRS) allow the collection of large amounts of traffic data [1]. For law enforcement agencies, this raises the problem of finding convenient techniques and tools to

M. Bernaschi · A. Celestini · S. Guarino · F. Lombardi (✉) · E. Mastrostefano
Institute for Applied Computing (IAC-CNR), Via dei Taurini 19, Rome, Italy
e-mail: flavio.lombardi@cnr.it

M. Bernaschi
e-mail: m.bernaschi@iac.cnr.it

A. Celestini
e-mail: a.celestini@iac.cnr.it

S. Guarino
e-mail: s.guarino@iac.cnr.it

E. Mastrostefano
e-mail: e.mastrostefano@iac.cnr.it

© Springer International Publishing AG, part of Springer Nature 2018
F. Leuzzi and S. Ferilli (eds.), *Traffic Mining Applied to Police
Activities*, Advances in Intelligent Systems and Computing 728,
https://doi.org/10.1007/978-3-319-75608-0_8

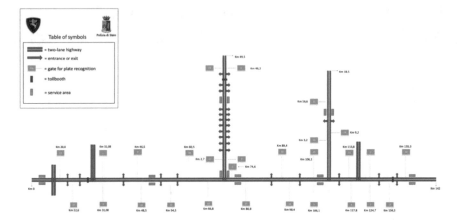

Fig. 1 The highway section under study

analyze such data, in order to find meaningful traffic patterns and identifying anomalous and criminal behaviors [2]. As a matter of fact, analyzing large amounts of traffic data is challenging due to the huge size of the dataset and the complexity of traffic dynamics. As such, developing an effective and scalable automatic traffic analysis system that can detect, track and gain useful insights about the behavior of road-users is vital to law enforcement agencies.

To support researchers willing to contribute to this difficult yet critical task, the Italian National Police (INP) made available a sample of data collected through automatic NPRS. The dataset consists of 365 Comma Separated Values (CSV) files,[1] each containing as many lines as the number of events registered by the 27 gates distributed along the highway section under study. Each line, in turn, contains the following fields: `plate`; `gate`; `lane`; `timestamp`; `nationality` (of the plate). Plates have been anonymized, mapping them to a set of consecutive positive integers. A sample of the data is shown below:

```
492788;20;1.0;2016-06-27 08:41:14;CH
144843;5;1.0;2016-06-27 12:38:13;I
144843;16;1.0;2016-06-27 09:01:30;I
7147369;4;2.0;2016-06-27 11:46:05;CH
```

Figure 1 shows the map provided along with the dataset. The map shows the highway section under study, where distances in Km between adjacent gates have been respected, but everything else has also been anonymized.

Hereafter, we describe the activities we carried out in order to analyze the sample of traffic data provided by the INP. Specifically, we present: (i) a set of exploratory bash/perl scripts, useful to compute descriptive statistics of the dataset and to

---

[1] The dataset consists of a file per day for the whole year 2016, but the file corresponding to October 7th is missing. As a consequence, the total number of files is 365 despite 2016 being a leap year.

highlight a few pitfalls it hides; (ii) a thorough python tool which, based on popular openly-available scientific libraries (numpy,[2] scipy,[3] matplotlib,[4] sklearn[5]), allows classifying both travel routes and plates. Remarkably, we used an unsupervised approach not relying on any a priori knowledge about the nature of the dataset or possible/interesting behaviors of road-users. Our findings show that relevant information can be extracted from similar transit records datasets despite the limitations introduced by the inevitable flaws of existing automatic NPRS. However, the analysis also suggests that directly applying standard statistical data mining techniques can hardly provide the desired insight into the intents of road-users. The unique characteristics of the problem require clustering algorithms to be composed and tailored based on a "measurable" definition of behavior of a vehicle. Simply guided by common sense, we therefore propose such a definition and highlight a few promising design choices (e.g., related to possible data filtering/cleaning strategies), which pave the way for future research in the area. Among the benefits of our approach, our classification mechanism provides significant information about the obtained clusters that can be rightfully expected to support law enforcement agencies in understanding/labelling the behavior of a road-user. In addition, our approach is inherently robust towards quality issues that are unavoidable with real datasets.

Roadmap. The paper is organized as follows: Sect. 2 reports a high-level statistical analysis of the dataset; Sect. 3 details the design of our classification/clustering tool; in Sect. 4 we summarize our findings; Sect. 5 discusses state-of-the-art approaches to traffic monitoring and analysis; finally, Sect. 6, draws conclusions and suggests possible directions for future work.

2   Statistical Analysis

We start with an overview of the information extracted from the Trap-2017 dataset by means of our bash/perl scripts.[6] Let us highlight that the procedures to extract information are pretty simple (few lines of scripting languages) but fully automatic. We do not discard a priori any data, including those that would appear clearly suspicious, for the simple reason that we do not look at the data but we directly use them as input to the scripts. Further, discarding data could prevent malicious behaviour from being discovered at a later analysis stage. This choice is motivated by the idea that our approach has to be scalable to much larger datasets where direct inspection is out-of-question.

---

[2]http://www.numpy.org/.

[3]https://www.scipy.org/.

[4]https://matplotlib.org/.

[5]http://scikit-learn.org/stable/.

[6]Details about the scripts and their output format can be found online http://twin.iac.rm.cnr.it/manuale.tbz.

A first statistics concerns the number of transits *per* plate. In fact, about 52% of the plates appear just once in the dataset. The total number of plates in the dataset is 14351059, whereas those for which we have records of at least two transits (i.e., at least one travel) are "only" 6958429. This immediately suggests that considering all available transits is not necessarily a good idea, since it means keeping in a lot of data that are not really informative of the behavior of any plate. Along this line, in view of extracting individual statistics for each plate we selected a relevant subset of the plates (26000), i.e. those that have more associated events. For each of them, we performed several analyses including, but not limited to: (i) collecting all events for that plate; (ii) identifying trips for each plate, where a trip is a sequence of transits separated by one hour at most; (iii) finding the speed for each pair of transits for each plate.

By looking at the extracted information (e.g., velocities or number of transits) it is apparent that there are either several unexpected findings or many errors in the data (or both...). For instance, there are 103897 events for the plate 257. This is a huge number corresponding to almost 300 passages *per* day. This is one of the anomalies that can be fed to domain experts. There are, obviously, possible alternative explanations of the phenomenon, such as: (i) the plate has been cloned and there is more than one vehicle using the same plate; (ii) the system that recognizes the plates may be wrong; (iii) the anonymization system may be wrong. Unfortunately, understanding what is the most reasonable explanation is not trivial. We cannot exclude that the *same* problems are present also for plates that have much smaller numbers of events. Despite the simplicity of the dataset it appears difficult to tell apart good and bad data.

Other significant issues emerge analyzing the correlation between two or among more than two events. We defined a trip as a sequence of two or more events in which two transits are separated by less than 3600 s. The motivation is that we aimed at identifying events that provided a reasonable proof that the considered vehicle had left the highway, so as to be able to describe the behavior of a vehicle in terms of its individual journeys, instead of considering its whole transits sequence altogether. Unfortunately, some of the journeys defined according to this convention are very strange (in a way that does *not* depend on our discretionary definition of trip). Here we mention just two of the anomalies we found by simply browsing the outputs of our scripts:

- Unrealistic velocities, such as velocities well beyond 1000 Km/h.
- Incoherent sequences of gates passages, such as several cases in which there is a transit through two gates that are not consecutive with no passage through the intermediate gate(s), even in sectors with no intermediate exits (e.g., transits through gates 2 and 6 with no passage through gate 7).

These kinds of anomalies also admit several possible explanations: (i) as before, these events may not belong, actually, to the same plate; (ii) the map that describes the highway sector may contain errors; (iii) our definition of journey could be deeply wrong (and any combination of them). In general, without detailed information about how the dataset was generated, making any assumption on what to expect seems hazardous.

In fact, our approach allows providing feedback to domain experts, even in a very early phase of the analysis. Nevertheless, additional findings (especially related to anomaly detection) to be fed to domain experts are discussed below.

3 Design of a Plates Behavior Classifier

As motivated in the Introduction, the challenge with the Trap-2017 dataset is to develop (semi-)automated classification systems able to identify road-users behaving in an unusual and potentially criminal way. In this Section we present a plates behavior classifier based on unsupervised clustering and accompanied by a cluster characterization to be used for understanding and labeling the clusters. We start with an overview of the underlying logic of the proposed tool, followed by a step-by-step description of its functioning.

3.1 Overview

Despite the INP suggested a few examples of recognizable suspect behaviors (cloned plates causing space-time inconsistent transits; habitual criminals visiting many service areas to select a victim), defining a complete model of malicious transit patterns is prohibitive. Unsupervised clustering is the only possible approach to prevent that unspecified and unpredictable/novel criminal behaviors are incorrectly classified, or not even recognized as malicious. However, rendering the final classification informative to police officers/technicians is vital in order to allow them to correctly interpret and label the obtained clusters.

Aiming at a precise, comprehensible and comparable/measurable definition of behavior of a plate, we introduce the concept of "route" to denote *any* possible pair of gates, and we move the focus of the analysis from gates transit records to *routes travel times*. We set aside rigid sequences of time-space coordinates in favour of a more flexible representation that enables inter-plate comparisons (e.g., to tell apart plates that visited a specific service area) while still allowing for the detection of inconsistencies. We do not limit the definition to pairs composed of two gates which are adjacent on the highway because, as emerged in Sect. 2, the dataset contains too many cases of consecutive close-in-time transits of a plate at two non-adjacent gates, which cannot be left out of the analysis. Since gates readings are not 100% reliable, even understanding when a plate exited or entered the highway is problematic. We therefore follow the sounder approach of using no filter and feeding our classifier with all available data. Eventually, the classifier will turn out to be able to do the job for us, demonstrating an excellent accuracy in detecting features of a route such as whether it is delimited by two adjacent gates.

Finally, as we deal with more complex tasks comparing what happened at different times may not be enough, and we may as well need to compare what happened at different (but somewhat matching) routes. For instance, identifying which plates stop "too often" at service areas requires elaborating on the travel times of different plates along all routes that contain a service area. We therefore need our plate behavior representation to embed a suitable classification of travel routes. The idea is to first rely on global travel times statistics to automatically classify all possible routes, and then cluster plates according to how their individual statistics fit the global statistics for each class of routes. Exactly as for plates, we discard any a priori classification of routes for at least two reasons: it may not work properly with an intrinsically flawed dataset, and it would affect the flexibility and scalability of the proposed solution.

Let us acknowledge that for the moment we set aside both the *lane* and *nationality* fields of each record, other than the time and date in which an event took place. Although we provided a few hints to motivate our choice, we do recognize that these data may have a relevance, and we expect using labels such as "preferred lane", "nationality", or "time and date" to be a potential boost to the performance of our clustering algorithms. Yet, investigating this option is left to future work.

3.2   The Tool

From the user viewpoint, our plates classification tool consists in a python script fed by a text file. Each line of the input file describes a different plate using the following syntax: $G_1 T_1 G_2 T_2 G_3 \ldots G_n$, i.e., a sequence of gates $G_i, \ldots, G_n$ separated by the corresponding travel times $T_i, \ldots, T_{n-1}$. For $1 \le i < n$, T_i is the travel time between gate G_i and gate G_{i+1}, expressed in milliseconds, and this syntax tells us that the first record involving that plate reports a transit at gate G_1, the second one reports a transit at gate G_2 exactly T_1 milliseconds after, and so on. Details about the usage of the tool and of all most relevant options can be found online.[7]

Step 0: Data Structures and Preprocessing. As mentioned before, our plates clustering algorithm relies on a suitable classification of routes, in turn based on travel times statistics. The step 0 of our tool is therefore the definition of a data structure which allows easy access to all travel times measured over each possible route, i.e., pair of gates. This data structure is created every time the tool is executed. Prior to compute the routes clustering, the tool allows applying a few transformations to the data, such as:

- Apply the same function to all travel times. For instance, applying a logarithmic transformation to all data may be used to make the distance between a few seconds and a few minutes comparable to the distance between a few minutes and a few hours.

---

[7]See http://twin.iac.rm.cnr.it/manuale.tbz.

- Normalizing all travel times relative to a specific route, dividing them by either their median or their mean. This is mostly useful to force easily comparable values for different travels, which may be critical for the correctness of the customized metrics we introduce later.
- Removing all times larger than a specific threshold. This threshold may be useful to prevent that pathological cases (such as broken down vehicle or cars that surely took an exit) poison data.

Step 1: Routes Clustering. The underlying idea for classifying the routes is that the distribution of travel times for a route can be rightfully expected to exhibit a few modes, corresponding to as many possible trends for drivers travelling along that route. For instance, looking at the distribution of travel times for a route we expect to find a peak at the expected travel time when the route is free, another peak at the expected travel time when the route is congested, and another peak at the expected travel time of drivers stopping at a service area (if there is at least one along that route). Comparing the coordinates of these peaks, we can estimate the similarity between two routes. Practically, this is done by applying Gaussian Mixture Modeling (GMM) [3], extending the aforementioned data structure by associating to each route its inferred Gaussian Mixture (GM). The distance between any two GMs is then established relying on a combination of the Earth Mover Distance (EMD) [4] and the Kullback-Leibler (KL) divergence [5]: KL is used to estimate the distance between any possible pair of Gaussians taken from the two mixtures (KL has a closed form for Gaussians), and these distances are fed to EMD together with the weights of the GMs. In short, our custom distance measures the cost of transforming the two GMs in one another, with the cost of transforming two Gaussians being their KL divergence. Once we have a distance matrix for all routes, we apply Hierarchical Agglomerative Clustering (HAC) [6] to find the desired routes classification. Let us remark that each class[8] can be described by its mean GM, which summarizes the expected travel time statistics of each member of that class.

Step 2: Plates Clustering. Let N be the number of routes clusters and M be the number of Gaussians in the GM modeling the travel times statistics of each route. Plates are classified using K-means clustering, associating a $N \times M$ matrix to each plate and using the norm of the difference of their matrices as the distance between two plates. The i, j element of the matrix associated to a plate is the fraction of data for that plate which refers to a route belonging to cluster i and that are assumed to have been generated by the Gaussian j associated to that route. In other words, the matrix associated to a plate summarizes the available information about the behavior of that plate in terms of a travel time distribution per cluster, and the more two plates have a similar behavior the closer their matrices are expected to be. In order to have comparable matrices we need M and N to be fixed for all routes. Additionally, having

---

[8]To avoid excessive repetitions we will interchangeably use "cluster" and "class" (sometimes even "type") to denote the partitions obtained using our classifier.

a fixed M is consistent with the use of an EMD-based metrics. We experimentally found $M = N = 5$ to be a good choice, while the optimal parameter K is found at each run relying on a silhouette score.

4   Our Findings

The main outputs produced at each run of our tool can be summarized as follows:

- For each possible pair of gates, which we call a *route*, the tool identifies the Gaussian Mixture (GM) that best fits the measured travel times on that route. Being, *de facto*, a generative model for transits over that route, this GM is a synthetic yet descriptive representation of behaviors on that route.
- The tool produces a classification of routes into clusters based on their GMs. This classification is accompanied by an aggregate GM which tells us what are the communal/distinctive features of that set of routes.
- Finally, the tool clusters plates based on their behavior, defined as the distribution over the Gaussians of each of the identified classes of routes. This behavior is also combined on a *per*-class level to obtain a single cumulative distribution describing the characteristics of each class.

4.1   Tuning the Classifier

Despite the ultimate goal of our tool is to classify plates, in order to measure the quality of our tool it is fundamental to also evaluated the results of the routes classifier. Indeed, while we do not have any prior knowledge about the plates, we can use the map provided along with the Trap-2017 dataset as a source of information to evaluate the obtained routes clusters obtained. Specifically, since we aim at identifying the options/parameters that provide a more accurate classification, we define two reasonable classes of routes and evaluate whether our classifier is able to recognize them. The two classes are:

- **Adjacent-Gates Routes (AGR)**: routes composed by two gates which are adjacent on the map. For instance, route (27,9) is in AGR, while route (27,26) is not.
- **No-Exit Routes (NER)**: routes which are compliant with the direction of travel, and can be therefore travelled without exiting the highway. For instance, route (27,26) is in NER, while (27,4) is not.

Of course, AGR is a subset of NER.

To measure how the accuracy of our route classifier is affected by different options we rely on the well known *Precision* (P) and *Recall* (R) scores, computing P and R for each cluster for both AGR and NER. Ideally, finding a cluster with both large P and large R means being able to recognize routes of the considered class (either AGR

Table 1 Accuracy of our route classifier

	Cluster 0		Cluster 1		Cluster 2		Cluster 3		Cluster 4	
	P	R	P	R	P	R	P	R	P	R
AGR	**1.00**	**0.72**	**1.00**	0.03	0.00	0.00	0.03	0.07	0.15	0.17
NER	**1.00**	0.12	**1.00**	0.01	0.07	0.24	**1.00**	**0.43**	**1.00**	0.20

(a) Raw travel time

	Cluster 0		Cluster 1		Cluster 2		Cluster 3		Cluster 4	
	P	R	P	R	P	R	P	R	P	R
AGR	0.00	0.10	0.00	0.00	**1.00**	**0.90**	0.00	0.00	0.00	0.00
NER	0.21	**0.85**	0.00	0.00	**1.00**	0.15	0.00	0.00	0.00	0.00

(b) Travel times ≤ 10 h

	Cluster 0		Cluster 1		Cluster 2		Cluster 3		Cluster 4	
	P	R	P	R	P	R	P	R	P	R
AGR	0.00	0.00	0.33	0.03	0.03	0.10	0.19	0.17	**0.87**	**0.67**
NER	0.04	0.12	0.33	0.01	**1.00**	**0.58**	**1.00**	0.15	**1.00**	0.14

(c) Normalized travel times

	Cluster 0		Cluster 1		Cluster 2		Cluster 3		Cluster 4	
	P	R	P	R	P	R	P	R	P	R
AGR	**0.60**	**0.93**	0.00	0.00	0.02	0.07	0.00	0.00	0.00	0.00
NER	**1.00**	0.27	0.00	0.00	**0.99**	**0.62**	0.03	0.11	0.00	0.00

(d) Logarithm travel times

or NER) with only a few false positives and false negatives. In general, the existence of one or more clusters with large P means that our classifier was able to detect the difference between the considered class and all other routes, whereas the existence of a single cluster with large R means that it was able to recognize the similarity among routes of the considered class.

In Table 1 we report the results of a preliminary set of experiments aimed at assessing the effects of four possible alternatives for data preprocessing: (a) doing nothing; (b) considering only travel times not larger than 10 h to remove episodic/irrelevant events; (c) using the median of travel times of a route as a "normalizing constant" to remove the dependence on the route length; (d) taking the logarithm of all travel times to adjust their density. The information gained by Table 1 can be summarized as follows:

- When using no preprocessing at all, AGR is nicely captured by Cluster 0, although a non negligible 28% of AGR routes are scattered over other clusters. The many clusters with P = 1 for NER mean that NER is split into many sub-classes, the largest of which contains only 43% (R = 0.43) of all NER routes.
- Cutting out all travel times larger than 10 h provides an excellent classification of AGR, with Cluster 2 having optimal P = 1 and R = 0.9. Unfortunately, NER is

not equally well clustered, since 85% of NER routes belong to Cluster 0 but they only sum up to 21% of the members of that cluster.

- Normalizing, a choice apparently consistent with the underlying logic of our metrics, yields Cluster 4 having $P = 0.87$ and $R = 0.67$ for AGR, and Cluster 3 having $P = 1$ and $R = 0.58$ for NER. However, non negligible percentages (10–17%) of each of the two classes are associated with other clusters.
- If we take the log of all travel times, an option which impacts on the GMM algorithm by modifying temporal distances, we see that Cluster 0 has $P = 0.60$ and $R = 0.93$ for AGR, while Cluster 2 has $P = 0.99$ and $R = 0.62$ for NER. Interestingly, Cluster 0 also has $P = 1$ and $R = 0.27$ for NER, meaning that the sub-optimal P of Cluster 0 for AGR and the sub-optimal R of Cluster 2 for NER are both due to part of NER routes being associated with AGR routes and thus being mapped to Cluster 0 instead of Cluster 2.

As we aimed at having two clusters respectively representing AGR and NER, in the following we will focus on option (d), i.e., taking the logarithm of all travel times. Indeed, we believe it is reasonable to associate the shortest NER routes to AGR, and we therefore pick the classification produced with option (d) as the neatest one. However, we point out that this choice is somewhat discretionary, and other preprocessing options could be preferred. For instance, a more detailed analysis of the map could lead to the conclusion that the classification provided by option (a) is a more refined partitioning of NER into meaningful sub-classes. Any deeper investigation is however left to future work.

4.2 Classifying Routes

Let us now focus on the ability of our classifier to model and cluster routes. As mentioned before, all following results refer to logarithmic preprocessing.

Figure 2 shows the travel time statistics, equipped with the inferred GM, for nine different routes. Routes (11,19), (24,17), (2,7) and (7,6) belong to the AGR class, whereas the other five routes are "anomalous" for some reason: routes (4,26) and (3,1) connect two gates located at the same Km of the highway, but on opposite directions; route (27,1) connects two gates which are both far away and on opposite directions; route (26,9) connects two adjacent gates, but on the wrong direction of travel; finally, route (9,9) is a loop.

Looking at the plots, a few observation can be made:

- Reasonable and anomalous routes can be easily told apart, since the former have a peak at a few minutes, while the latter are unbalanced towards a day or a week. This difference is well captured by their respective GMs.
- Among reasonable routes, we can clearly distinguish routes (11,19) and (2,7) from routes (24,17) and (7,6), despite all four routes are in AGR and contain exactly one service area. The main divergence between these two pairs is that (11,19) and (2,7) are around 10 Km long, compared with (24,17) and (7,6) being more than 40

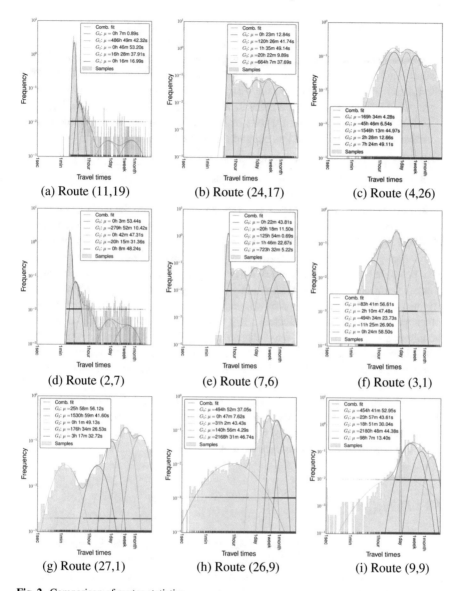

Fig. 2 Comparison of routes statistics

Km long, a difference which impacts on both the mean time and the variance of behaviors of road-users on these routes and explains why (24,17) and (7,6) are the only two routes of AGR associated to routes Cluster 2. Notably, routes (24,17) and (7,6), which correspond to the very same highway section but in the two opposite direction of travel, have a almost identical GM.

- Among anomalous routes, (3,1) and (4,26) can be easily paired and distinguished from the others, which is desirable since they are two completely analogous routes, as previously described. In general, the behaviors captured by the GMs of these routes are somewhat expectable, except for route (27,1) having a non-negligible Gaussian with mean time less than 2 min. This is completely unrealistic as the shortest feasible path between gates 27 and 1 is surely several tenths of Km long. This aspect may be the proof of one or more cloned plates, and it would require further investigation which lies beyond the scope of this paper.

Finally, let us focus on the routes clusters identified by our classifier. Table 2 reports a synthetic description of these clusters, where for each of them we report its size and a aggregate GM obtained averaging the GMs of all the members of the cluster. For these aggregate GMs, we report the mean (μ) and the weight (w) of each Gaussian, ordered by the latter. Clusters 1 and 4 are composed by just a few routes. Significantly, all but one of these routes are loops, i.e., routes delimited on both sides by the same gate, but even considering both clusters together we only get about a third of all possible loops. All in all, these two clusters seem of scarce interest. Conversely, Clusters 0 and 2, as already emerged in Sect. 4.1, group together all routes that are compliant with the direction of travel, and the only significant difference between routes appearing in the two clusters seem to be their length. This is also visible in the aggregate GMs of these two clusters: Cluster 0 contains shorter routes whose travel times are strongly unbalanced towards values in the order of a few minutes, which is the expected travel time of a vehicle running across that route at fast but licit speed; on the other hand, along longer routes, such as those belonging to Cluster 2, we detect a flatter distribution, in which travel times around an hour or even half a day are notably common. Finally, Cluster 3 contains all other routes. As previously highlighted, for this larger class of routes much larger travel times, ranging from half a day to as much as two months, are extremely recurring.

4.3 Classifying Plates

Finally, we can switch to our plates classifier and assess what can be inferred about the behavior of different classes of plates based on its outcomes.

Table 3 presents a summary of the clusters obtained by our plates classifier. We recall that each plate is described by means of the distribution of its travel times over the aggregate GMs we produced for each routes cluster. Indeed, GMM *de facto* provides a generative model for travel times measured over a route, allowing each recorded event to be interpreted as an outcome of a more general behavior. Considering aggregate GMs for an entire routes cluster is a way to define a restricted set of possible behaviors that may occur over all routes of that class. Now, plates can be classified based on how frequently they behave in a specific way over routes of a specific type. In Table 3, other than the size of each cluster, we report highlights of the typical behavior of plates belonging to that cluster, in terms of a short description

Table 2 Routes Clusters

*RC	*Size	G_0		G_1		G_2		G_3		G_4	
		μ	w	μ	w	μ	w	μ	w	μ	w
0	45	09 m 02 s	0.69	16 m 11 s	0.20	2 h 13 m 37 s	0.07	34 h 12 m 44 s	0.03	542 h 38 m 17 s	0.02
1	9	179 h 19 m 16 s	0.34	46 h 3 m 29 s	0.24	893 h 48 m 24 s	0.17	19 h 50 m 32 s	0.13	12 h 18 m 54 s	0.12
2	106	1 h 16 m 28 s	0.34	14 m 36 s	0.32	12 h 27 m 24 s	0.13	158 h 31 m 21 s	0.12	1226 h 24 m 23 s	0.09
3	568	258 h 50 m 18 s	0.30	60 h 37 m 2 s	0.26	1446 h 59 m 20 s	0.21	10 h 22 m 19 s	0.16	1 h 13 m 27 s	0.07
4	1	155 h 4 m 30 s	0.44	1233 h 35 m 15 s	0.28	1 h 22 m 20 s	0.17	23 h 25 m 20 s	0.09	1 s	0.01

Table 3 Plates Clusters

*PC	*Size	Behavior 1				Behavior 2				Behavior 3			
		RC	G	μ	w	RC	G	μ	w	RC	G	μ	w
0	2380633	3	2	60h37m2s	0.13	0	3	34h12m44s	0.12	3	4	1446h59m20s	0.12
1	635257	3	0	1h13m28s	0.47	3	1	10h22m19s	0.32	3	3	258h50m18s	0.03
2	3680336	0	0	9m2s	0.45	0	3	34h12m44s	0.16	0	4	542h38m17s	0.09
3	247398	3	3	258h50m18s	0.79	3	1	10h22m19s	0.04	3	2	60h37m2s	0.03

of the Gaussians they are most frequently associated with. For instance, the first quadruplet of Cluster 0 says that plates belonging to that cluster are associated 13% of the time to Gaussian 2 of Routes Cluster (RC) 3, which means that 13% of the events involving those plates are travels along routes of RC 3 each having average temporal length of 60 h 37 m 2 s.

A few significant insights can be deduced from a closer look at Table 3:

- Cluster 2 contains more than 3.5M plates which exhibit a very standard behavior: they almost exclusively travel over routes of type 0, which are the most reasonable type, and half of these travels occur at the most frequent/expected speed. We argue that these plates are vehicles driven by ordinary road-users, and thus they can be pruned from the dataset if the goal is identifying criminals. If our intuition is correct, this would mean reducing by more than a factor 2 the total number of plates to be classified.
- Cluster 0 is also very large, summing up to more than 2M plates. Contrarily to Cluster 2, however, plates belonging to this cluster exhibit a very diverse and odd behavior, which probably deserves a deeper investigation (e.g., through a second layer of clustering).
- Cluster 3 is composed of plates that, altogether, are associated almost 80% of the time with travel times of approximately 10 days over anomalous routes. A possible explanation is that these are plates for which we only have a few records, thus what we interpret as travels are actually occasions in which they left the highway to only come back a few days after.
- Finally, Cluster 1 is probably the most interesting one, since it is characterized by only two very recurring habits: travelling over routes of RC 3 with travel times either close to 1 h or to 10 h, which are both reasonable values despite implying two different activities. We do not have enough information to really understand the meaning of similar behaviors. However, it would not be surprising to discover they can be associated with known suspicious/criminal patterns, such as moving goods from one point of the highway to another.

5 Related Work

5.1 Traffic Monitoring and Analysis

A large amount of work on traffic monitoring and analysis has been carried out in the past. In fact, making sense and extracting meaningful information from the large data flows collected by cameras and GPS tracking is not a trivial task.

Sivaraman's et al. work [1] is an interesting starting point over on-road vision-based vehicle detection, tracking, and behavior understanding. They provide a survey of recent works in the literature, placing vision-based vehicle detection in the context of sensor-based on-road surround analysis. Authors detail advances in vehicle detection, discussing monocular, stereo vision, and active sensor-vision fusion for on-road vehicle detection. Authors also characterize on-road behavior, introducing common performance metrics and benchmarks. Our present work is much different, as we are not in control of the way data is captured. However, we have tried to infer a posteriori some information from the classification of the given raw data. Interesting work on traffic pattern analysis and optimization can be found in a work by Koller et al. [2]. They leverage machine vision-based technology and high-level symbolic reasoning to develop a system for detailed, reliable traffic scene analysis. Their symbolic reasoning approach uses a dynamic belief network to make inferences about traffic events such as vehicle lane changes and stalls. Koller's work is complementary to ours, and interesting future work can be foreseen by integrating mutual results.

5.2   Pattern Mining and Clusterization

Jindal et al. [7] analyze the problem of mining frequent patterns from road traffic data by developing a method to mine spatiotemporal periodic patterns in the traffic data and use these periodic behaviors to summarize the huge road network. Their first step is to find periodic patterns from the speed data of individual road sensor stations, then use their periods to represent the station's periodic behavior using probability distribution matrices. Jindal uses density-based clustering to cluster the sensors on the road network based on the similarities between their periodic behavior as well as their geographical distance, thus combining similar nodes to form a road network with larger but fewer nodes. This work is somewhat similar to our present work. However, our approach is somehow more universal as it can be applied to heterogeneous data with little or no prerequisites. Elfeky et al. [8] use periodicity mining to predicting trends in time series data. They address the problem of detecting the periodicity rate of a time series database. They define different types of periodicities and propose scalable algorithms performing in $O(n \log n)$ time for a time series of length n. Also Kiran et al. [9] aim at discovering partial periodic itemsets in temporal databases. They introduce a new measure (periodic-frequency) to determine the periodic interestingness of itemsets by taking into account their number of cyclic repetitions in the entire data. These two contributions are interesting and can be leveraged for a given analysis following our approach described here.

Giannotti et al. [10] leverage a knowledge discovery process to mine frequent travel patterns, big attractors and extraordinary events influence on mobility. They aimed at predicting dense traffic areas in the near future. They also defined M-Atlas, a querying and mining language that eases the analytical process by transforming raw GPS tracks into mobility knowledge. Giannotti's work will be considered for future work in combination to our present results. Necula [11] performed R-based statistical analysis to identify contiguous set of road segments and time intervals which have the largest statistically significant relevance in forming traffic patterns. He mined vehicle traces to extract outlier traffic patterns. Similarly to what we did, he organized the road infrastructure as segments in a graph and tracks the visits for each vehicle. He found that over time, the visited segments settle into a pattern and vary periodically. Grossi et al. [12] address the problem of clustering data as a machine learning problem, as well as an optimization problem. They present a constraint programming model for a centroid based clustering and one for a density based clustering. In particular, as a key contribution, they show how the formulation of the density-based clustering by constraint programming makes it very similar to the label propagation problem and they propose a variant of the standard label propagation approach. Their approach is quite different from the one presented here. Future work will investigate benefits and pitfalls of our approach in comparison to Grossi's.

6 Conclusions and Future Work

In this paper we have described the efforts, pitfalls, and successes of applying a purely automated classification/clustering approaches to analyze the TRAP-2017 challenge dataset. All our work was performed leveraging open source tools, self-written (python, perl, bash) code and state of the art scientific software libraries. Various approaches to data filtering/cleaning have been manually applied and compared, and all obtained results and figures have been analyzed and discussed. Our findings show that unsupervised clustering is a viable approach to extract meaningful information about the composition of the dataset. Further, by building our classifier upon a formal and descriptive definition of behavior of a plate, we have created the conditions for police officers to fully characterize the clusters produced by our tool. Additional analysis and understanding of the results will be part of future work. We believe the results described here can pave the way to interesting research on the matter.

References

1. Sivaraman, S., and M.M. Trivedi. 2013. Looking at vehicles on the road: A survey of vision-based vehicle detection, tracking, and behavior analysis. *IEEE Transactions on Intelligent Transportation Systems* 14 (4): 1773–1795.
2. Koller, D., J. Weber, T. Huang, J. Malik, G. Ogasawara, B. Rao, and S. Russell. 1994. Towards robust automatic traffic scene analysis in real-time. *Proceedings of the 33rd IEEE Conference on Decision and Control*, 3776–3781.
3. McLachlan, Geoffrey, and David Peel. 2004. *Finite mixture models*. New York: Wiley.
4. Rubner, Yossi, Carlo Tomasi, and Leonidas J. Guibas. 2000. The earth mover's distance as a metric for image retrieval. *International Journal of Computer Vision* 40 (2): 99–121.
5. Kullback, Solomon, and Richard A. Leibler. 1951. On information and sufficiency. *The Annals of Mathematical Statistics* 22 (1): 79–86.
6. Joe, H. 1963. Ward Jr. Hierarchical grouping to optimize an objective function. *Journal of the American Statistical Association* 58 (301): 236–244.
7. Tanvi Jindal, Prasanna Giridhar, Lu An Tang, Jun Li, and Jiawei Han. 2013. Spatiotemporal periodical pattern mining in traffic data.

8. Elfeky, Mohamed G, Walid G Aref, and Ahmed K Elmagarmid. 2005. Periodicity detection in time series databases. *IEEE Transactions on Knowledge and Data Engineering* 17 (7): 875–887.
9. Uday Kiran, R., Haichuan Shang, Masashi Toyoda, and Masaru Kitsuregawa. 2017. Discovering partial periodic itemsets in temporal databases. *Proceedings of the 29th International Conference on Scientific and Statistical Database Management,SSDBM '17*, 30:1–30:6. New York: ACM.
10. Giannotti, Fosca, Mirco Nanni, Dino Pedreschi, Fabio Pinelli, Chiara Renso, Salvatore Rinzivillo, et al. 2011. Unveiling the complexity of human mobility by querying and mining massive trajectory data. *The VLDB Journal* 20 (5): 695–719.
11. Emilian Necula. Analyzing Traffic Patterns on Street Segments Based on GPS Data Using R. Transportation Research Procedia, 10:276–285. 2015. *18th Euro Working Group on Transportation, EWGT 2015, 14–16 July 2015*. The Netherlands: Delft.
12. Grossi, Valerio, Anna Monreale, Mirco Nanni, Dino Pedreschi, Franco Turini, et al. 2015. Clustering formulation using constraint optimization. *Selected Papers of SEFM 2015 Workshop on Software Engineering and Formal Methods*, 93–107. New York: Springer.

Vehicle Classification Based on Convolutional Networks Applied to FMCW Radar Signals

Samuele Capobianco, Luca Facheris, Fabrizio Cuccoli and Simone Marinai

Abstract This paper investigates the processing of Frequency-Modulated Continuous-Wave (FMCW) radar signals for vehicle classification. In the last years, deep learning has gained interest in several scientific fields and signal processing is not one exception. In this work we address the recognition of the vehicle category using a Convolutional Neural Network (CNN) applied to range-Doppler signatures. The developed system first transforms the 1-dimensional signal into a 3-dimensional signal that is subsequently used as input to the CNN. When using the trained model to predict the vehicle category, we obtained good performance.

1 Introduction

The automatic processing of real-time traffic information is an important approach to analyze the traffic. The number of traffic sensors is quickly growing, as well as the amount of useful information available for traffic monitoring applications [1]. As we are moving towards Smart Cities, traffic monitoring becomes an important topic to address in order to improve the safety of traveling and to help the police monitoring work.

The previously mentioned features are required to build an intelligent transportation system where one important part is vehicle detection and classification. Vehicle

S. Capobianco (✉) · L. Facheris · S. Marinai
Università degli studi di Firenze, via Santa Marta 3, Firenze, Italy
e-mail: samuele.capobianco@unifi.it

L. Facheris
e-mail: luca.facheris@unifi.it

S. Marinai
e-mail: simone.marinai@unifi.it

F. Cuccoli
CNIT RaSS c/o Dipartimento di Ingegneria dell'Informazione,
via Santa Marta 3, Firenze, Italy
e-mail: fabrizio.cuccoli@unifi.it

© Springer International Publishing AG, part of Springer Nature 2018
F. Leuzzi and S. Ferilli (eds.), *Traffic Mining Applied to Police
Activities*, Advances in Intelligent Systems and Computing 728,
https://doi.org/10.1007/978-3-319-75608-0_9

classification is an important task also for police activities in order to prevent potential criminal behaviors.

Many solutions have been adopted to detect or classify vehicles based on different types of sensors [2–4]. The traffic sensors produce a large quantity of data which can be useful as input to various machine learning techniques useful to address the vehicle detection and classification tasks. Specifically, in this work we use deep learning to analyze signals coming from radars.

Deep learning addresses neural network architectures that are composed by several transformation layers which learn representations of the input data. These architectures define multiple levels of abstraction learning different representations from the row data, irrespective of application contexts. These methods have dramatically improved the state-of-the-art in speech recognition, visual object recognition, and many other application research fields.

One example of detection vehicles from camera videos is proposed for the automatic car counting. Using the OverFeat [5] framework (Convolutional Neural Network and SVM) and the Background Subtraction Method, the authors have demonstrated to be able to count cars. Another solution for a low-cost vehicle detection and classification system is based on a FMCW radar [6]. The radar, unlike video sensors, is less vulnerable to adverse weather conditions and is able to work with the same performances in any light condition. In particular, the latter is a remarkable feature to build a reliable traffic control system.

In this paper, we address the vehicle classification in a highway context by using information coming from a continuous wave radar. The Italian Traffic Law classifies the vehicles in six categories which can run on the highways. It is important to check the speed of running vehicles because each category has different speed limit. At present, the continuous wave radar is used by the major Italian highway company to monitor the speed of traveling motor vehicles on highway. We want to discover the right vehicle category using the radar sensor signal already employed to check the speed limit in order to help the police monitoring.

The rest of the paper is organized as follows. In Sect. 2 we describe the main features of the radar that provides the data used for vehicle classification. In Sect. 3 we analyze the overall system organization and the neural network architecture. The experiments performed are presented in Sect. 4, while concluding remarks are in the conclusions (Sect. 5).

2   Signal Model of FMCW Radar

In this work, we use signals collected by a 24 GHz Frequency-Modulated Continuous Wave radar in order to classify vehicles running on highways. As sketched in Fig. 1, the radar was placed on a fixed structure at 5.3 m height with the antenna illuminating a single lane and pointing with a depression angle of 32° toward the back of the vehicles passing under the structure along that lane.

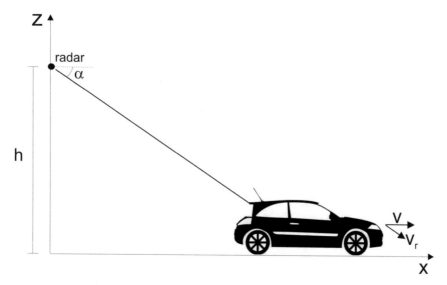

Fig. 1 Scheme of the acquisition geometry of the FMCW radar (h = 5.3 m, $\alpha = 32°$)

As mentioned, data employed in this work come from a FMCW radar. Like pulse Doppler radars, FMCW radars can measure range and velocity of targets (vehicles in our case). However, the continuous signal transmission brings to the designer the benefit to reduce significantly the transmission power needed to transmit the same energy of a pulsed system, with a significant positive impact also on their size and cost. Furthermore, transmitter and receiver operate continuously, independently and simultaneously so that the receiver does not need to be silenced during the transmission. For the reasons briefly explained in the following, the price paid for these significant advantages is the need for a quite higher computational load with respect to pulsed radars. Nowadays, however, this is not a problem anymore: the availability of low cost, small size and high speed programmable FPGAs on the market is the reason why FMCW radars are widely employed in a great number of applications, ranging from remote sensing to law enforcement and automotive industry. The waveform transmitted by a FMCW radar is a frequency modulated signal:

$$A \cdot cos \left(2\pi f_0 t + 2\pi \int_{-\infty}^{t} m(\alpha) d\alpha \right) \tag{1}$$

where the modulating signal $m(t)$ - coinciding with the instantaneous frequency deviation of the modulated signal - is a periodic triangular waveform with period $2T$

$$m(t) = \Delta f \cdot \sum_{n=-\infty}^{\infty} \left[Tri \left(\frac{t - (n+1)T}{T} \right) - \frac{1}{2} \right] \tag{2}$$

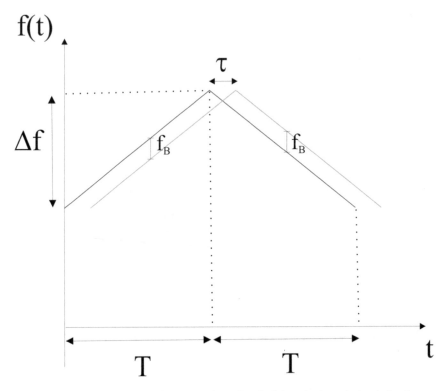

Fig. 2 Instantaneous frequency of the transmit (red) and of the echo (green) signals for the case of absence of relative motion between the radar and the target. Only two ramps are plotted for the sake of clarity

with

$$Tri(t) = \begin{cases} 1+t & \text{if } -1 < t \leq 0 \\ 1-t & \text{if } 0 < t < 1 \\ 0 & \text{elsewhere} \end{cases} \tag{3}$$

and where Δf, that coincides in practice with the bandwidth of the modulated signal, is the variation of its instantaneous frequency during the so called "sweep interval" T. Note that, during two subsequent sweep intervals, the variation of the instantaneous frequency is of opposite sign: if it increases during the first interval (ascending or up-ramp) at a rate $\Delta f/T$, in the second interval (descending or down-ramp) it decreases of the same quantity and at the same rate. If the relative radial velocity between the radar and the target is zero, we have the situation sketched in Fig. 2: the instantaneous frequency of the received (echo) signal (in green) will be the delayed replica of the transmitted signal (in red), the delay τ being equal to 2R/c, where R is the range of the target and c the speed of light.

The received signal is converted to baseband by beating it with the reference transmit signal with a mixer. This generates a sinusoidal signal with the same frequency f_B (beat frequency) in both the up- and down-ramp case, and that is linearly related to the delay of the echo signal as follows

$$\tau = \frac{T}{\Delta f} \cdot f_B \qquad (4)$$

which allows to measure the range of the target. Therefore, in FMCW systems the range scale corresponds to a frequency (beat) scale, while in pulsed systems it corresponds to a "simpler" time (delay) scale. This implies that a spectral analysis is needed in the FMCW case, with the consequent higher computational load mentioned above. Consider in fact the case of several stationary targets: each of them will give rise to echo signals of different intensity that, beaten with the reference transmit signals, will generate sinusoidal contributions with different frequencies (and amplitudes) at baseband. Intensive spectral (FFT) processing is therefore needed to detect targets and to resolve them in range. A further complication arises from the presence of a relative motion between radar and target. In such a case, assuming a relative radial velocity v_r, the received signal frequency is affected by a Doppler shift equal to $f_D = \pm \frac{2v_r}{\lambda}$, where λ is the wavelength and the sign depends on the direction of the relative motion. In Fig. 3 sketches the situation: the Doppler frequency is negative, so it could represent the case of Fig. 1. In this case we get two different beat frequencies for the up- and down-ramps:

$$f_{B1} = \frac{\Delta f}{T} \cdot \tau + |f_D| \quad ; \quad f_{B2} = \frac{\Delta f}{T} \cdot \tau - |f_D| \qquad (5)$$

from which we can derive both delay and Doppler frequency of the target, and consequently its range and radial velocity. In the presence of multiple targets, determining their parameters with accuracy may ask for careful spectral processing.

In the case under exam, the interest is focused only on one target (the vehicle passing under the structure on which the radar is mounted). However, the objective is not only to determine its velocity, but also to classify the vehicle, which would be a challenge also for pulsed radars due to the number of parameters involved (speed, size, height, length, backscattering mechanisms involved and backscattered power) and to the possibility of artifacts (e.g. two different vehicles running close to each other). The 24 GHz radar parameters used for measurements are $\Delta f =120$ MHz and T = 40 ms. It was decided to base the classification procedure on spectrograms obtained by 512 points FFT processing applied to the echo signals corresponding to the up- and down-ramps.

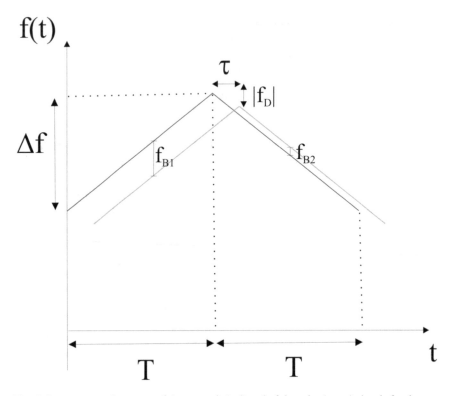

Fig. 3 Instantaneous frequency of the transmit (red) and of the echo (green) signals for the case of presence of relative motion between the radar and the target. Only two ramps are plotted for the sake of clarity

3 Application Scenario

Along the highway infrastructures are placed radar sensors used to monitor the vehicles speed by law enforcement to prevent potential criminal behaviors. In this work, we use the signal from FMCW radar to classify the vehicles in order to replace the camera sensors typically used for this task.

3.1 Signal Processing

The radar system used for monitoring the vehicles produces a beat frequency corresponding to each observed vehicle running along a lane. The signal is a sequence of frames whose duration depends on the time taken by the vehicle to travel along the antenna footprint in the observed lane. As illustrated in the upper part of Fig. 4 (time-domain signal), the radar echo signal generated at baseband after having beaten

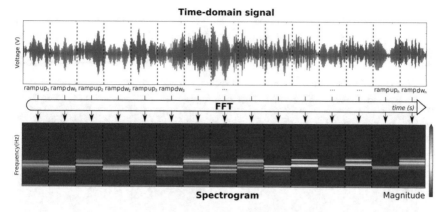

Fig. 4 The received FMCW radar signal is beaten with the transmitted signal: in the upper part of the figure is shown an example of the time domain signal obtained in that manner. The duration of the time slots (windows) shown in the upper picture is that of the ramps (up and down ramps) of the transmitted signals. Fast Fourier Transform (FFT) is then applied to each window. The two spectrograms that we utilized for further processing were derived by separating the upper ramp windows of the signal from the down ramp ones. The spectrogram corresponding to the up ramp was obtained by building a matrix whose columns are the FFT moduli of the sequence of up ramp windows. The spectrogram corresponding to the down ramp was obtained analogously. In the lower picture of the figure, the level of the computed modulus grows according to the color bar at the right of the lower picture

the received signal with the transmitted one, is a sequence of received signals corresponding to the alternate sequences of transmitted up- and down-ramps.

One of the typical ways to analyze the radar beat frequency through time is to compute the so called range-Doppler signature. Such an approach has been already used to detect vehicles in a similar context [6]. Using the Short Time Fourier Transform (STFT) we are able to carry out the time-frequency analysis. STFT is a well known technique in signal processing to analyze non-stationary signals in the time domain. Given a non-stationary signal, a moving window of fixed duration selects a sequence of signal segments and computes a sequence of Fast Fourier Transform (FFT) on them. Windows may overlap or not for reasons related to optimal spectrum estimation. This procedure allows to analyze the spectral properties of the non stationary signal during their variations over time. Figure 4 depicts the process we used to compute the spectrograms from the original signals. Considering one of our FMCW radar signals and using a moving window (without overlap) with a duration equal to the ramp duration (T), we extracted two sequences of alternate signal segments. For each segment we computed the FFT modulus that was then used to build the spectrogram, the first one corresponding to the sequence of up-ramp signals, the second one to that of down-ramp signals. Figure 5 shows the baseband signal collected from a vehicle and the two spectrograms obtained by applying the STFT algorithm. These two representations describe each vehicle running along the observed lane.

The idea to classify the running vehicles on the highway is to learn directly from the range-Doppler signature inherent to each spectrogram a data representation useful

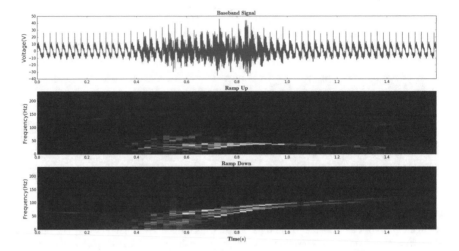

Fig. 5 Radar signals. Top frame: baseband signal obtained by beating the received signal with the transmitted signal generated by one running vehicle observed by a FMCW radar. In the middle and in the bottom frames are respectively shown the range-Doppler signature for the up- and down-ramp

to discriminate the categories. In this case, the time-frequency analysis transforms a 1-dimensional signal (top signal in Fig. 5) into two 2-dimensional signals (up-ramp and down-ramp in Fig. 5). Since we have a representation similar to that of an image, it is possible to use a neural network based on Convolutional Neural Network architecture to capture the topological structure of the input spectrograms and use the learned representation to classify the signals.

3.2 DeepRadarNet

In the last years, Convolutional Neural Networks obtained very good results in several different tasks. These architectures are used for Object Recognition tasks [7–9] improving also the accuracy score with respect to human performance [10]. These architectures require very large datasets to learn all their parameters. For instance, one of the most popular datasets used to train these networks is the Imagenet [11], that contains 3.2 millions of images according to the WordNet hierarchy.

Convolutional Neural Networks (CNNs) are a particular type of artificial neural network consisting of alternated convolutional and spatial pooling layers [12]. The convolutional layers generate feature maps by linear convolutional filters followed by nonlinear activation functions (e.g. rectifier, sigmoid, tanh). In 2012 the AlexNet [13] CNN architecture outperformed other machine learning techniques which were based on hand-crafted features in the ImageNet Large Scale Visual Recognition Challenge.

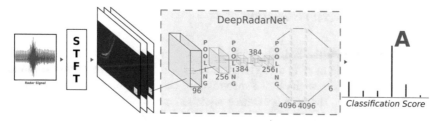

Fig. 6 Given a 1-dimensional radar signal, using the Short Time Fourier Transformation we are able to compute three range Doppler signatures (up-ramp, down-ramp, average-ramp). Then, these three computed channels are stacked into one tensor. A trained *DeepRadarNet* model maps the tensor into the classification score predicting the class of vehicle

Our proposed architecture, named *DeepRadarNet*, is inspired to the AlexNet architecture. We depict the model architecture in Fig. 6.

The model consists of eight main layers (five convolutional and three fully connected layers). The first convolutional layer has 96 kernels with a receptive field of $3 \times 11 \times 11$ size and a stride of 4 units. The output of the first layer is then pooled to the first max-pooling layer where the max pooling and response normalization has been performed. The second layer has 256 kernels with a receptive field of $96 \times 5 \times 5$ size that is followed by another max-pooling layer. The max-pooling computes a max operator with kernel 3×3 and stride 2×2. The third, fourth and fifth convolutional layers have a different number of kernels and receptive field dimensions. Third layer: 384 kernels $256 \times 3 \times 3$ receptive field; fourth layer: 384 kernels and $384 \times 3 \times 3$ receptive field: fifth layer 256 kernels and $384 \times 3 \times 3$ receptive fields. The last convolution layer is followed by the last max-pooling. The computed features are the input for three fully connected layers which map the computed features into the output neurons. The sixth and the seventh layer compute a fully-connected transformation and each layer is composed by 4096 neurons. To avoid overfitting, the dropout [14] layer is applied between the previous fully-connected layers. The eighth layer corresponds to the output neurons that match the number of classes.

The output of the last fully-connected layer is fed into an N-way softmax which models a distribution over the N class labels. Learning the model parameters is an optimization task which can be solved using multi class cross-entropy cost function used during the training phase.

In this work we consider a classification task with six vehicle categories, where each category is associated to an output neuron. Using this architecture, we can classify the input signal into vehicle categories defined according to the Italian Traffic Law.

3.3   Overall System Structure

Given a signal from the FMCW radar, we can compute two spectrograms which describe the vehicle running along an observed segment of highway lane. After the STFT transformation, we have two spectrograms (2-dimensional signals) as range-Doppler signature. The length of these signals depends on the vehicle speed. In order to have the same dimension for each observation, we need to uniform the input shape. To this purpose, we fix the input shape according to the sample size distribution. Through zero-padding, we adjust the computed spectrograms obtaining a tensor composed by three channels: 1) the up-ramp, 2) the down-ramp and 3) the average between the two previous channels. We therefore have one tensor with a fixed size for each sample.

The three channels obtained in this manner are useful as input to the proposed neural network, which maps one sample to the classification score that correspond to the category of the observed vehicle. The input tensors are mean normalized using the train set average tensor. The mean compensation is applied both to the train and test set signal tensors.

After the training phase, we use the best trained model to predict the vehicle category. As shown in Fig. 6, the input signal goes through the STFT operator which computes three matrices tensors: one channel for each signal representation (up-ramp, down-ramp, average-ramp). After the zero-padding size normalization, the category is predicted from the trained *DeepRadarNet* model. The classification score is a way to define the vehicle category.

4   Experiments

In this section we describe the experiments carried out to evaluate the proposed solution. As detailed in Fig. 7, the dataset is composed by 9, 981 baseband signals not uniformly distributed in the six categories: car (A), car-trailer (B), truck (C), cargo truck (D), bus (E) and motorcycle (F).

As we have seen in Sect. 3.1 it is possible to compute the range-Doppler signature and visualize the spectrograms as shown in Fig. 8.

Each signal length depends on the vehicle speed during the observation, since the number of frames for each spectrogram is inversely proportional to the vehicle speed. We therefore need to normalize this input size to a fixed size in order to use the signals as input to the model.

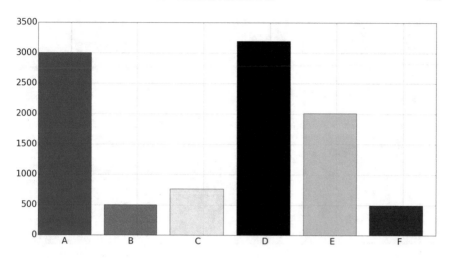

Fig. 7 Distribution of the 9, 981 samples into the 6 classes. We can see as the classes are not uniformly distributed. In particular, we have more examples for car, cargo truck and bus categories (A, D and E classes)

4.1 10-Fold Cross Validation

To assess the ability of the system to discriminate the vehicle categories from the range Doppler signature we use the k-fold cross validation. In this case we extract randomly 10 different folds, each composed by 2400 samples for training, 270 samples for validation and 7311 samples for testing. In particular, the train and validation set have 400 samples and 45 samples for each category, respectively. We shuffled the dataset before extracting each fold.

During the training phase we use the stochastic gradient descent to adjust the model parameters. We use one batch size with one sample for each category (six samples for each batch). The learning rate, momentum, and weight decay are respectively set to 0.0001, 0.9 and 0.0005.

One of the problems for training deep neural networks is the parameters initialization, especially when the training set is not large enough. One possible solution is to transfer the learned weights from another application context [15]. In our case we initialize the convolution layer parameters using the Alexnet pre-trained model [13], exception made for the last fully connected layers, that are randomly initialized.

Figure 9 shows the confusion matrix for one folder and the average results for all the folders. Note that the class F (motorcycle) obtains a prediction accuracy of 1 which corresponds to a perfect classification score.

For the classes A, B, and D which are car, car-trailer, and cargo truck respectively, we obtain good results because the network is able to properly describe the shape related to their spectrograms. For these three classes we can see in Fig. 8 some examples from the dataset and we can get as these spectrograms have discriminative shapes though a large portion of them are quite similar to each other.

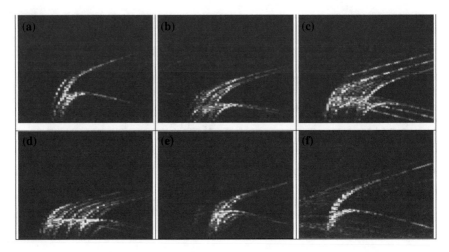

Fig. 8 One transformed example for each category. We represent the range-Doppler signature as the average signal between the up- and the down-ramp. The meaning of the letters is the following: **a** car, **b** car-trailer, **c** truck, **d** cargo truck, **e** bus, **f** motorcycle

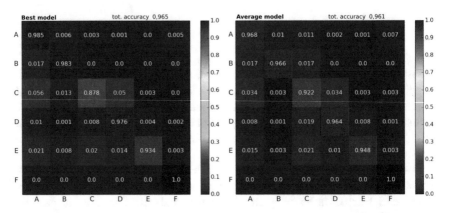

Fig. 9 The class confusion matrix of the results obtained. The average recognition accuracy is 0.961 considering ten different folds. On the left the confusion matrix for one of the best folders. On the right the average results considering all folders

For the classes C and E, which are truck and bus respectively, we have some misclassification due to the related spectrogram shapes. Always in Fig. 8, if we consider the vehicle structure (captured by the radar signal), the categories C and E include different shapes which are more difficult to characterize and classify.

5 Conclusions

In this preliminary work we wanted to investigate the capability of Convolutional Neural Networks to recognize vehicle categories by means of a FMCW radar signal. In particular, we have utilized a technique to transform a 1-dimensional signal to a 3-dimensional tensor based on Short Time Fourier Transformation. The computed tensors have then been used to train our convolutional architecture named *DeepRadarNet*.

Using the transfer learning technique, it is possible to initialize the model weights moving the learned parameters from a pre-trained model for Object Recognition task to our *DeepRadarNet* model. In this way, the training phase starts from a better starting point, improving also the recognition performances. We obtained good results in this prediction task, which encouraged us to continue the research in this direction.

In the future work we want to address also the vehicle detection directly on the radar signal, trying to improve the performance in this vehicle classification task. In particular, we want to compare different deep learning architectures considering also Recurrent Neural Networks. The idea is to compare such architectures so as to evaluate the system performance in terms of accuracy and of memory consumption. The final target is to develop a stand alone low-cost system able to work on cheaper single-board computer.

Acknowledgements The authors wish to tank Infomobility S.R.L. Concordia sulla Secchia (Modena, Italy) and Autostrade per l'Italia (Roma, Italy) for having provided the radar data.

References

1. Munoz-Ferreras, J.M., Calvo-Gallego, J., and Perez-Martinez, F. 2008. Monitoring road traffic with a high resolution lfmcw radar. In *IEEE Radar Conference*, 1–5.
2. Ki, Y.K., and D.K. Baik. 2006. Vehicle-classification algorithm for single-loop detectors using neural networks. *IEEE Transactions on Vehicular Technology* 55 (6): 1704–1711.
3. De Angelis, G., A. De Angelis, V. Pasku, A. Moschitta, and P. Carbone. 2016 A simple magnetic signature vehicles detection and classification system for smart cities. In *IEEE International Symposium on Systems Engineering (ISSE 2016)*, 1–6.
4. H. Sandhawalia, J. A. Rodriguez-Serrano, H. Poirier, and G. Csurka. 2013. Vehicle type classification from laser scanner profiles: A benchmark of feature descriptors. In *16th International IEEE Conference on Intelligent Transportation Systems (ITSC 2013)*, 517–522.
5. Sermanet, Pierre, David Eigen, Xiang Zhang, Michaël Mathieu, Robert Fergus, and Yann Lecun. 2014. Overfeat: Integrated recognition, localization and detection using convolutional networks.
6. Fang, J., H. Meng, H. Zhang, and X. Wang. 2007. A low-cost vehicle detection and classification system based on unmodulated continuous-wave radar. In *2007 IEEE Intelligent Transportation Systems Conference*, 715–720.
7. Christian Szegedy, Wei Liu, Yangqing Jia, Pierre Sermanet, Scott Reed, Dragomir Anguelov, Dumitru Erhan, Vincent Vanhoucke, and Andrew Rabinovich. 2015. Going deeper with convolutions. In *CVPR 2015*.
8. K. Simonyan and A. Zisserman. 2014. Very deep convolutional networks for large-scale image recognition. *CoRR*, abs/1409.1556.

9. Jonathan Long, Evan Shelhamer, and Trevor Darrell. 2015. Fully convolutional networks for semantic segmentation. *CVPR (to appear)*.
10. Kaiming He, Xiangyu Zhang, Shaoqing Ren, and Jian Sun. Delving deep into rectifiers: Surpassing human-level performance on imagenet classification. In *Proceedings of the IEEE International Conference on Computer Vision (ICCV 2015)*, pages 1026–1034.
11. J. Deng, W. Dong, R. Socher, L.-J. Li, K. Li, and L. Fei-Fei. 2009. ImageNet: A Large-Scale Hierarchical Image Database. In *CVPR09*.
12. Yann Lecun, Léon Bottou, Yoshua Bengio, and Patrick Haffner. 1998. Gradient-based learning applied to document recognition. In *Proceedings of the IEEE*, 2278–2324.
13. Alex Krizhevsky, Ilya Sutskever, and Geoffrey E. Hinton. 2012. Imagenet classification with deep convolutional neural networks. In *Advances in Neural Information Processing Systems 25*, ed. F. Pereira, C.J.C. Burges, L. Bottou, and K.Q. Weinberger, 1097–1105. Curran Associates, Inc.
14. Srivastava, Nitish, Geoffrey Hinton, Alex Krizhevsky, Ilya Sutskever, and Ruslan Salakhutdinov. 2014. Dropout: A simple way to prevent neural networks from overfitting. *Journal of Machine Learning Research* 15: 1929–1958.
15. Jason Yosinski, Jeff Clune, Yoshua Bengio, and Hod Lipson. 2014. How transferable are features in deep neural networks? In *Advances in Neural Information Processing Systems 27*, ed. Z. Ghahramani, M. Welling, C. Cortes, N.d. Lawrence, and K.q. Weinberger, 3320–3328. Curran Associates, Inc.

Traffic Data: Exploratory Data Analysis with Apache Accumulo

Massimo Bernaschi, Alessandro Celestini, Stefano Guarino,
Flavio Lombardi and Enrico Mastrostefano

Abstract The amount of traffic data collected by automatic number plate reading systems constantly incrseases. It is therefore important, for law enforcement agencies, to find convenient techniques and tools to analyze such data. In this paper we propose a scalable and fully automated procedure leveraging the Apache Accumulo technology that allows an effective importing and processing of traffic data. We discuss preliminary results obtained by using our application for the analysis of a dataset containing real traffic data provided by the Italian National Police. We believe the results described here can pave the way to further interesting research on the matter.

1 Introduction

The widespread availability of smart devices capable of collecting and producing large amounts of data poses the problem of how to store and process such data. In particular, the issue mainly concerns how to make such data valuable and how to extract meaningful information from them. The term Big Data [2, 5] is often used in this context, albeit its definition is not fully agreed upon by researchers of both industry and academia. Big Data definition can slightly vary on the basis of the area in which the term is used. In general, it refers to large datasets that cannot be managed and processed by traditional software/hardware tools within an

M. Bernaschi · A. Celestini · S. Guarino · F. Lombardi (✉) · E. Mastrostefano
Institute for Applied Computing (IAC-CNR), Via dei Taurini 19, Rome, Italy
e-mail: flavio.lombardi@cnr.it

M. Bernaschi
e-mail: m.bernaschi@iac.cnr.it

A. Celestini
e-mail: a.celestini@iac.cnr.it

S. Guarino
e-mail: s.guarino@iac.cnr.it

E. Mastrostefano
e-mail: e.mastrostefano@iac.cnr.it

© Springer International Publishing AG, part of Springer Nature 2018
F. Leuzzi and S. Ferilli (eds.), *Traffic Mining Applied to Police
Activities*, Advances in Intelligent Systems and Computing 728,
https://doi.org/10.1007/978-3-319-75608-0_10

acceptable time frame. In this case, we deal with an excerpt of a big dataset containing traffic data, that grows continuously and rapidly over time. Data are collected by automatic number plate reading systems deployed on a limited segment of an Italian highway. As such, it seemed interesting to study a data analysis solution that could scale to larger datasets, providing the possibility to analyze the whole dataset with our approach. By leveraging the highly scalable Accumulo database, we propose a scalable, fully automated procedure to import and process data, while extracting features, items and events from the dataset. Present paper aims at describing the efforts, pitfalls, and findings of applying such a scalable approach to analyze real traffic data, provided by the Italian National Police. We believe the results described here can pave the way to further interesting research on the matter.

Our contributions can be summarized as follows: (a) we introduce an automated approach, based on the highly-scalable Accumulo database, to the processing of the dataset; (b) we reason over the results and feasibility of the proposed approach; (c) we provide a working implementation of our solution, that has been already deployed on a Virtual Machine submitted to the *shared task* proposed in the announcement of the conference.

The rest of the paper is organized as follows: Sect. 2 shortly introduces Apache Accumulo; Sect. 3 discusses results of exploratory data analysis performed on the dataset; Sect. 4 describes the Accumulo application and its usage; Sect. 5 discusses state of the art approaches for traffic monitoring and analysis; finally, Sect. 6, draws conclusions and suggests possible directions for future work.

2 Apache Accumulo

Apache Accumulo[1] [3] is a key-value database modeled after Google's Bigtable [1]. Accumulo is highly scalable, distributed and open source, it stores its data on the Apache Hadoop's HDFS[2] filesystem and uses Apache Zookeeper[3] for consensus. Accumulo was developed with a focus on building analytical applications and it is able to horizontally scale across thousands of machines. Accumulo features automatic load-balancing and partitioning, data compression and fine-grained security labels.

Data Model: Accumulo is a key-value store in which keys are kept sorted at all times in ascending byte-by-byte lexicographical order (i.e. roughly in alphabetical order). Key-value pairs are organized in tables which in turn are partitioned in tablets. Tablets are automatically distributed across multiple machines in the cluster. Accumulo keys are made up of several fields providing a richer data model than simple key-value databases (see Table 1). In particular, the key is composed by three main

---

[1] http://accumulo.apache.org.

[2] http://hadoop.apache.org.

[3] https://zookeeper.apache.org.

Table 1 Accumulo key-value data model

Key					Value
Row ID	Column			Timestamp	
	Family	Qualifier	Visibility		

fields named: Row ID, Column and Timestamp. Each column is in turn split into three components: Column Family, Column Qualifier and Column Visibility. The Row ID field is used to logically group several key-value pairs, all pairs with same Row ID are considered as part of the same row. The Timestamp field allows storing more than one version of the same Value. Usually, this field is automatically filled while data are imported. By default Accumulo keeps only the newest version of a key-value pair. The Column Visibility field is used to provide fine-grained access control to key-value pairs, allowing to protect data stored in the same physical cluster and with different sensitivity levels.

Accumulo is written in Java and provides a Java client library. It provides also a Thrift[4] proxy that can be used to write client applications in other languages. All elements of the Key and the Value are represented as byte arrays except for the Timestamp field, which is a Long. The default constraint on the maximum size of the Key is 1 MB.

Motivation: We wanted to investigate the feasibility and potentialities of a data analysis based on table stores that are are emerging as a viable alternative to traditional relational databases. In particular, we selected Apache Accumulo considering also benchmarks results available in the Literature [8, 12, 13]. Such studies report a pretty good performance for ingest and query rate, and show Accumulo scalability properties over clusters of different sizes. We were also interested in its distinctive features, namely the server-side programming framework called iterators[5] and the fine-grain data access control provided by cell-level security.

3 Data Analysis with Accumulo

The dataset provided by the Italian National Police contains traffic data, along with a map of the street segment where data have been collected. Both the dataset and the map have been anonymized; plates are identified by a numerical ID and the map contains only information about relative distances in Km between consecutive gates. Specifically, the dataset contains a set of CSV-like files in which columns are separated by a semicolon. Each file's row contains information about the

---

[4]http://thrift.apache.org.

[5]Other Bigtable systems have integrated server-side programming, e.g. Apache HBase provides coprocessors, but they are not so deeply and nicely integrated in the underlying technology as iterators [3].

Fig. 1 Conceptual data
model

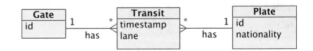

transit of a plate, and each row is composed by the following five columns: `plate;` `gate;` `lane;` `timestamp;` `nationality`. The *plate* column contains the plate numerical ID; the *gate* column contains the gate number crossed by the plate; the *lane* column contains the lane number traversed by the plate during the transit; the *timestamp* column contains the date of transit; the *nationality* column contains the nationality associated to the plate.

Starting from the given dataset, we developed the conceptual model shown in Fig. 1. We identify three main entities: plate, gate and transit. Each plate is identified by a number and a nationality and it can cross several gates. Each gate is identified by a number and it can be crossed by several plates. Finally, a transit is identified by a timestamp, a lane, a plate number and a gate number. As shown in Fig. 1, the transit entity acts as connection between plate and gate entities.

To analyze the dataset, we designed and implemented an Accumulo-based application. In the following section we comment on preliminary results we obtained using our application. In Sect. 4 we better describe our application and its usage.

3.1   Data Processing

In order to store and process the dataset, we created two Accumulo tables named trap_plate and trap_gate respectively, whose structure is summarized in Tables 2 and 3. Trap_plate is used to store data and statistics regarding each and every plate present in the dataset, whereas trap_gate is used to store data and statistics

Table 2 Schema of trap_plate table

Plate number	GATE	Transit date	List of gates
	LANE_NUMBER	Lane number	Number of transits
	GATE_NUMBER	Gate number	Number of transits
	NATION	Nation name	number of transits
	TOT_TRANSITS		number of transits

Table 3 Schema of trap_gate table

Gate number	TRANSIT	Transit date	List of plates
	PLATE	Plate number	Number of transits
	TOT_TRANSITS		Number of transits

regarding each gate. In particular, for each plate we store: the number of transits (column TOT_TRANSITS), the nationalities assigned to the plate and their frequency (column NATION), the frequency of each gate crossed by the plate (column GATE_NUMBER), the list of gates crossed with the transit timestamp (column GATE), the lanes traversed and their frequency (column LANE_NUMBER). For each gate we store: the number of transits (column TOT_TRANSITS), the list of plates that crossed the gate with the transit timestamp (column TRANSIT), the list of plates that crossed the gate and their frequency (column PLATE). The application is designed to store data and compute statistics at the same time, we leverage the Accumulo iterator to accomplish that task. Iterators are part of the server-side programming framework provided by Accumulo. They can be used to customize the behavior of tables. Iterators are basically simple functions that add logic to a subset of key-value pairs. Accumulo applies iterators in succession and their order is defined by a priority value. The output of each iterator is a set of sorted key-value pairs that is used as input by the following iterator. Accumulo supplies a set of built-in iterators, but it is possible to create custom ones. In our application, we use the built-in `SummingCombiner` iterator and the custom `ListArrayCombiner` iterator. The former interprets Values as Longs and returns the sum of the set of values, the latter interprets Values as array of Longs and returns a single array of Longs containing all the values. The `SummingCombiner` iterator is applied to column families TOT_TRANSITS, NATION, GATE_NUMBER and LANE_NUMBER of table trap_plate, and to column families TOT_TRANSITS and PLATE of table trap_gate. The `ListArrayCombiner` iterator is applied to column family GATE of table trap_plate and column family TRANSIT of table trap_gate.

3.2 Data Cleaning and Exploratory Data Analysis

Once data are processed and stored inside Accumulo's tables, we can perform some checks on the dataset, looking for anomalies, i.e. evidence of different kinds of inconsistencies and/or violations of the conceptual model shown in Fig. 1. Such process would possibly highlight errors in the raw data, that can be removed for later analysis. We can then proceed exploring and analysing the dataset to better understand its contents.

For what concerns anomalies, we focus on two main cases:

Multiple Nationalities: plates with multiple nationalities;
Simultaneous Transits: plates with simultaneous transits on the same gate or on different gates.

The former case aims at identifying plates to which more than one nationality has been associated, violating the "one nationality per plate" constraint. The latter case aims at identifying plates with space-time inconsistencies, e.g. plates simultaneously crossing the same or different gates. It is worth noticing that 7392630 plates of the dataset (i.e. approximately 51.5% of all plates) have only one transit. Such plates are clearly excluded by our analysis because we don't have enough information about them, as discussed in the following.

Multiple Nationalities. We check the presence of plates to which more than one nationality has been associated, i.e. CSV files' rows with the same values in plate column but with different values in the nationality column. The dataset allows the presence of question marks '?' in the nationality column. We assume that in those cases the system was unable to assign a nationality to the plate. In our analysis, we distinguish between two cases to which we assign two different level of severity:

- plates with two or more valid nationalities and possibly a question mark; [red severity]
- plates with one valid nationality and a question mark. [yellow severity]

To identify such plates in the dataset, we use the trap_plate table (see Sect. 3.1). All the information we need is stored in the NATION column.

In the dataset we found 112255 plates corresponding to a yellow severity and 5464 plates corresponding to a red severity, summing up to 117719 global anomalies. An excerpt of plates with multiple nationalities is reported in Table 4 and 5. The dataset contains 14351059 plates, thus anomalies represent approximately the 0.78% and the 0.03% respectively of all plates. As previously reported, 51.5% of all plates occur only once in the dataset. As such, we chose to exclude such plates from our analysis, given that (at most) they can have one associated nationality or a question mark. Plates with at least 2 transits are 6958429. In this case anomalies represent approximately the 1.61% and the 0.07% respectively of the plates in the dataset.

Simultaneous Transits. We check the presence of plates with simultaneous transits, i.e. CSV files' rows with same values in columns plate and timestamp. In our analysis we distinguish between two cases to which we assign two different level of severity:

Table 4 Plates with one valid nationality and a question mark. For each nationality/question mark the frequency is reported after the colon

Yellow severity	
Plate	Nationalities
239	?:555 – MP:4238
240	?:1093 – MP:9887
247	?:1066 – MP:24302
257	?:5142 – MP:98755
14347760	?:2 – RO:3
14348522	?:4 – H:1
66473	?:415 – I:103
99101	?:10 – CH:1
99430	?:3 – H:381
100647	?:8 – SLO:400

Table 5 Plates with two or more valid nationalities and possibly a question mark. For each nationality/question mark the frequency is reported after the colon

Red severity	
Plate	Nationalities
23409	CH:1 – I:2
28465	HR:1 – I:329
28466	HR:5 – I:716
110835	?:1 – HR:21 – I:202
301624	?:1 – CH:2 – I:3
422605	CH:30 – I:6
394818	HR:34 – I:153
2411956	?:11 – HR:103 – I:38
14106861	HR:1 – I:11
14106869	?:1 – HR:3 – I:3

- plates with simultaneous transits on the same gate; [yellow severity]
- plates with simultaneous transits on different gates. [red severity]

To identify such plates in the dataset we use the trap_plate table (see Sect. 3.1), all the information we need is stored in the column GATE. In the dataset we found 231961 transits corresponding to a yellow severity and 780 transits corresponding to a red severity, i.e. 232741 anomalies overall. An excerpt of plates with simultaneous transits is reported in Tables 6a, b. The number of different plates for yellow and red severity is 105928 and 61 respectively, corresponding approximately to the 0.73% and the 0.0004% of all plates. If, as in the previous case, we consider only plates with at least 2 transits, the anomalies represent approximately 1.52% and 0.0008% of that set, respectively.

Plates Transits. We extract and visualize some statistics about plates' transits. In particular, Fig. 2a, b show the number of (annual) transits for each gate. Indeed, we notice that transits are not equally distributed among the gates. As such, we can easily identify which gates are more visited or less visited. These figures give us a first information about traffic flaws and plates' behaviors on the highway segment covered by the dataset.

Figure 3a, b show the empirical distribution of plates' transits, suggesting that plates' transits approximately follow a power-law distribution. From the figures we observe that there is a majority of plates that appears only few times in the dataset, i.e. they pass through the highway segment sporadically, and a small subset of plates that produce almost all traffic data. This latter set of plates has a very high transit frequency, and is composed of few plates.

Plates Transits on Pairs of Gates. We analyze plates' transits on consecutive gates, i.e. plates' transits on pairs of gates (PoG). We consider any pair of gates appearing in the dataset, not only those that are valid based on the map provided along with the data. In the dataset there are 27 different gates, and it is interesting to notice that

Table 6 Simultaneous Transits

<table>
<tr><td colspan="3" align="center">Yellow Severity</td><td colspan="3" align="center">Red Severity</td></tr>
</table>

Plate	Gates	Timestamp
10	14 - 14	2016-08-03 11:55:43
25	13 - 13	2016-07-11 15:40:19
25	1 - 1	2016-08-03 08:15:57
37	1 - 1	2016-08-19 07:26:48
39	15 - 15	2016-01-13 15:50:38
382	27 - 27	2016-12-23 12:16:21
259	1 - 1 - 1 - 1	2016-08-29 15:04:02
20991	18 - 18	2016-08-03 00:08:13
8735	4 - 4	2016-08-03 07:15:43
20294	13 - 13	2016-08-03 07:39:21

Plate	Gates	Timestamp
239	4 - 8	2016-12-23 11:13:10
240	20 - 15	2016-02-17 07:01:45
247	5 - 13	2016-12-27 13:31:34
257	14 - 21	2016-01-04 12:36:20
119729	26 - 9	2016-06-06 13:58:55
119729	10 - 13	2016-06-24 14:13:28
8536	2 - 16	2016-11-15 11:37:43
8614	10 - 18	2016-04-13 15:41:40
2267	18 - 7	016-12-07 08:21:58
267	18 - 10	2016-12-07 10:06:51

(a) Plates with simultaneous transits on the same gate

(b) Plates with simultaneous transits on different gates

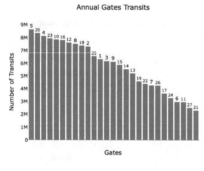

(a) Annual transits per gate

(b) Cumulative annual transits distribution

Fig. 2 Annual transits

any possible pair of gates appears at least once in the dataset. Figure 4 shows the number of transits for 50 PoG, an excerpt of all pairs that are 729 overall. There is no apparent cut on the transits' distribution, i.e. a point from which the number of transits decreases, highlighting a clear drop in the distribution. This would suggest the possibility to distinguish *"normal"* and *"anomalous"* pairs based on the number of transits. In our approach, *"normal"* pairs represent transits in the right direction and concern consecutive gates. Vice-versa *"anomalous"* pairs represent transit in the wrong direction, concern not consecutive gates or gates impossible to cross without crossing gates that are in-between them.

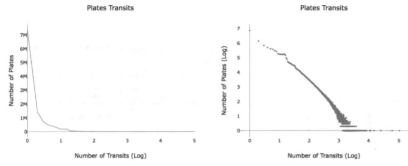

(a) Plates transits, x-axis is in log scale.

(b) Plates transits, x-axis and y-axis are in log scale.

Fig. 3 Plates transits statistics

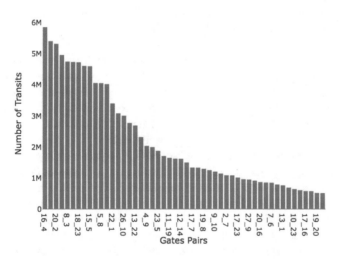

Fig. 4 Transits per gates pairs

The analysis of single PoG suggests the existence of common patterns concerning the time spent to cross pairs, at least for the *normal* ones. Figure 5a shows the crossing times distribution for two PoG, both normal. The maximum value of the x-axis in the chart is set to 30 minutes. In both cases we notice some peaks, two for the pair 26_10 and three for the pair 23_15, denoting different plates' behaviors while crossing the gates. The first peak of pair 26_10 corresponds to a speed of approximately 126 km/h, the second to a speed of approximately 86 km/h (speed can be computed using the map provided along with the data). The first peak of pair 23_15 corresponds to a speed of approximately 125 km/h, whereas the other two peaks correspond to a speed

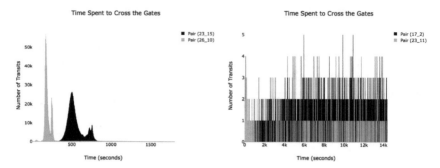

(a) Transits distribution of normal pairs of gates

(b) Transits distribution of anomalous pairs of gates

Fig. 5 Transits distribution per pairs of gates

of approximately 88 and 83 km/h respectively. Thus, for normal gate pairs, different plates' behaviors seem to emerge from the data.

Figure 5b shows the crossing times distribution for two PoG, both anomalous. The maximum value of the x-axis in the chart is set to 240 min. In this case, it doesn't seem to exist any clear pattern in the data. In particular, we observe a very low number of transits for these pairs, two orders of magnitude lower than the case of normal pairs shown in Fig. 5a. Moreover, transit times appear somehow equally distributed over time. Thus, in case of anomalous pairs, no plates' behaviors, denoting a preference on how to cross the PoG, seem to emerge from the data.

The two figures show just few example of PoG. Analyzing others pairs, we observe the same shapes in the transits' distribution, with a clear difference between anomalous and normal pairs. We also tried to compare pairs comprising or not comprising a service area, but it is not clear how to differentiate the two cases just looking at the charts (note that pair 26_10 has a service area, whereas pair 23_15 doesn't).

Finally, Fig. 6a, b show a graph representation of the highway segment under study. Each node of the graph represents a gate, while each edge represents a connection between consecutive gates. Edge size is proportional to the number of annual transits per pair of gates and edge color is proportional to the speed. In Fig. 6b we computed the mean speed while in Fig. 6a we computed the median, in both cases we considered only transits in a 1 h window.

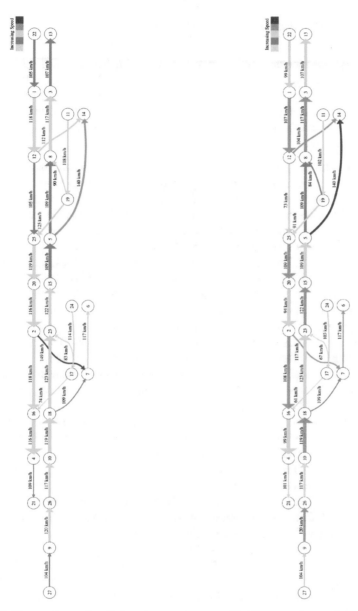

(a) Median speed, max transit time 1 hour

(b) Mean speed, max transit time 1 hour

Fig. 6 Number of transits and speed per pairs of gates

4   Application Usage

Hereafter, we describe the application we designed, implemented and used to produce the results discussed in Sect. 3. The application provides some scripts that perform various operations on the dataset. All scripts require Apache Hadoop's HDFS,
ZooKeeper and Accumulo. Some configuration parameters can be set through a
.properties file named config.properties (see Sect. 4.1).
   Main scripts are:

- **LoadDataset.sh**: This script imports data in the Accumulo database and processes them. It performs the following operations: it connects to Accumulo and
  creates all tables needed to import the dataset; it reads all the files contained in the
  dataset and writes the information in the Accumulo tables; finally it creates a file
  in the application home directory named dataset.load.stats containing
  summary information about importing operations.
- **CheckDataset.sh**: This script performs some checks on the imported dataset. In
  particular, it checks the presence of plates with multiple nationalities and/or simultaneous transits on gates (same or different). It creates five files in the directory
  specified by the parameter *dataFolder*, that can be set through the configuration
  file *config.properties*. The files created by the script are: nation.yellow.
  anomalies and nation.red.anomalies containing the list of plates with
  multiples nationalities; transits.yellow.anomalies and transits.
  red.anomalies containing the list of plates with simultaneous transits on
  gates; and plates.total.transits containing the distribution of plates'
  transits.
- **GatesTransits.sh**: This script extracts all transits for each gate creating a file
  named GateA.gate.transits, where GateA is a gate number. Each file row
  contains the information about a plates' transits on the gate, each row has two
  columns separated by a space. The first column contains the timestamp of the
  transit, the second column contains a list of plates separated by a colon. The script
  also creates a file named gates.off, containing time intervals for which a gate is
  considered off (see configuration parameter *gateOffTimeInterval*, Sect. 4.1). Each
  file row contains four columns separated by a space. The first column is the gate
  number, the second column is the starting date of the off period, the third column
  is the end date of the off period. The fourth column is the elapsed time between
  timestamps.
- **GatesPairsTransits.sh**: This script extracts all transits for each gates pair and
  the list of gates crossed by each plate. Specifically, it creates a file named
  plates.transits containing a row for each plate. Each file row contains
  the plate's number and a list of crossed gates. The gates list contains the time spent
  to cross each gates pair traversed by the plate. Moreover, the script creates for
  each gate pair a file named GateA_GateB.pair.transits where GateA and
  GateB are two gate numbers, possibly the same. Each file row contains the information about a plate's transits. Each row has four columns separated by a comma. The
  first column contains the time interval required by the plate to cross the two gates

(in milliseconds), the second column contains the date at which the plate crossed the first gate, the third column contains the date at which the plate crossed the second gate and the fourth column contains the plate number. Finally the script creates (for each gate pair) a file named `GateA_GateB.pair.transits.cleaned` where GateA and GateB are two gate numbers, possibly the same. These files have the same structure of the previous ones, but do not contain transits for time interval in which at least one gate is considered off. To create these files the script leverages data contained in the file `gates.off` created by *gatesTransits.sh*.

4.1 Configuration

The file config.properties is used to set configuration parameters of the application, available parameters are:

- **AccumuloUser**: the username used to access the Accumulo database.
- **AccumuloPwd**: the password used to access the Accumulo database.
- **AccumuloTable**: the prefix used to generate the name of each table.
- **AccumuloIstance**: the Accumulo instance to connect to.
- **ZookeeperServers**: the Zookeeper servers used by Accumulo.
- **DatasetFolder**: the folder in which the dataset (to import) is stored
- **DataFolder**: the folder in which all data (files) produced by the application will be stored.
- **nThreads**: the number of threads used to perform operations.
- **GateOffTimeInterval**: a threshold value used to decide when a gate is considered off. If a gate doesn't read any plate for a time interval bigger than gateOffTimeInterval, the gate is considered off for that interval.

5 Related Work

Traffic monitoring is a complex activity that may collect data through cameras, sensors and/or GPS tracking. A fairly large amount of work has been performed to improve traffic monitoring quality and analysis scalability in general. Some relevant examples are described in the following.

The state-of-the art review of the literature by Sivaraman and Trivedi [14] is a good starting point about on-road vision-based vehicle detection, tracking, and behavior understanding. They provide a survey of recent works in the literature, placing vision-based vehicle detection in the context of sensor-based on-road surround analysis. Authors detail advances in vehicle detection, discussing monocular, stereo vision, and active sensor-vision fusion for on-road vehicle detection. Authors also introduce spatiotemporal measurements, trajectories, and various features to characterize on-road behavior, introducing common performance metrics and benchmarks. Further interesting work on traffic pattern analysis and optimization can be found in a work by Koller et al. [10]. They leverage machine vision-based technology and high-level symbolic reasoning to develop a system for detailed, reliable traffic scene analysis. Their symbolic reasoning approach uses a dynamic belief network to make inferences about traffic events such as vehicle lane changes and stalls. Emilian Necula [11] applies a statistical approach on vehicle traces that are mined to extract outlier traffic patterns. He organizes the road infrastructure as segments in a graph and tracks the visits for each vehicle. Over time, the visited segments settle into a pattern and vary periodically. Necula performs statistical analysis using R to identify contiguous sets of road segments and time intervals which have the largest statistically significant relevance in forming traffic patterns.

Hoh and Gruteser [6] discuss automotive traffic monitoring using probe vehicles with GPSes. With a special attention to privacy concerns, they propose a system based on virtual trip lines and an associated cloaking technique. Virtual trip lines are geographic markers that indicate where vehicles should provide location updates. These markers can be placed to avoid particularly privacy sensitive locations. They also allow aggregating and cloaking several location updates based on trip line iden- tifiers, without knowing the actual geographic locations of these trip lines. Thus they facilitate the design of a distributed architecture, where no single entity has a com- plete knowledge of probe identities and fine-grained location information. Knospe et al. [9] present a detailed analysis of single-vehicle data to analyse microscopic inter- action of the vehicles. Their analysis of free flow and synchronized traffic provides information about wide jams which persist for a long time. Elfeky et al. [4] analyze periodicity mining for predicting trends in time series data. They address the problem of detecting the periodicity rate of a time series database. Two types of periodicities are defined, and a scalable, computationally efficient algorithm is proposed for each type. The algorithms perform in $O(n \log n)$ time for a time series of length n.

An example of leveraging Big Data technology for traffic analysis is provided by Keller el al. [7] describing a system for combining heterogeneous (air) traffic management data using semantic integration techniques transforming data into a unified semantic representation within an ontology-based triple store. This is an interesting issue and their solution can potentially be applied to automotive traffic analysis scenarios.

6 Conclusions and Future Work

In this paper we have described the efforts, achievements and issues of applying a scalable approach to analyze real traffic data, provided by the Italian National Police. All work was carried out at IAC-CNR leveraging Apache Accumulo database and self-written Python and Java code. Various approaches to data filtering, cleaning and analysis have been applied together with a purely automated scalable NoSQL database approach. In fact, the Accumulo database was used to allow fast and scalable data access in order to extract interesting features from the dataset. We believe the results described here can pave the way to further interesting research in the future.

References

1. Chang, Fay, Jeffrey Dean, Sanjay Ghemawat, Wilson C. Hsieh, D.A. Wallach, M. Burrows, T. Chandra, A. Fikes, and R.E. Gruber. 2008. Bigtable: A distributed storage system for structured data. *ACM Transactions on Computer Systems* 26 (2): 4:1–4:26.
2. Chen, Min, Shiwen Mao, and Yunhao Liu. 2014. Big data: A survey. *Mobile Networks and Applications* 19 (2): 171–209.
3. Cordova, Aaron, Billie Rinaldi, and Michale Wall. 2015. *Accumulo: Application development, table design, and best practices*, 1st ed. Beijing: O'Reilly Media, Inc.
4. Elfeky, Mohamed G., Walid G. Aref, and Ahmed K. Elmagarmid. 2005. Periodicity detection in time series databases. *IEEE Transactions on Knowledge and Data Engineering* 17 (7): 875–887.
5. Erl, Thomas, Wajid Khattak, and Paul Buhler. 2016. *Big data fundamentals: Concepts, drivers & techniques*. Boston: Prentice Hall Press.
6. Hoh, Baik, Marco Gruteser, Ryan Herring, Jeff Ban, Daniel Work, Juan-Carlos Herrera, Alexander M. Bayen, Murali Annavaram, and Quinn Jacobson. 2008. Virtual trip lines for distributed privacy-preserving traffic monitoring. In *Proceedings of the 6th international conference on Mobile Systems, Applications, and Services*, MobiSys '08, 15–28. USA: ACM.
7. Keller, Richard, M., Shubha Ranjan, Mei, Y. Wei, and Michelle, M. Eshow. 2016. Semantic representation and scale-up of integrated air traffic management data. *Proceedings of the International Workshop on Semantic Big Data SBD '16*, 4:1–4:6. USA: ACM.
8. Kepner, J., W. Arcand, D. Bestor, B. Bergeron, C. Byun, V. Gadepally, M. Hubbell, P. Michaleas, J. Mullen, A. Prout, A. Reuther, A. Rosa, and C. Yee. 2014. Achieving 100,000,000 database inserts per second using accumulo and d4m. In *2014 IEEE High Performance Extreme Computing Conference (HPEC)*, 1–6.
9. Knospe, W., L. Santen, A. Schadschneider, and M. Schreckenberg. 2002. Single-vehicle data of highway traffic: Microscopic description of traffic phases. *Physical Review Series E* 65: 056133.
10. Koller, D., J. Weber, T. Huang, J. Malik, G. Ogasawara, B. Rao, and S. Russell. 1994. Towards robust automatic traffic scene analysis in real-time. In *Proceedings of the 33rd IEEE Conference on Decision and Control*, 4, 3776–3781.
11. Necula, Emilian. 2015. Analyzing traffic patterns on street segments based on gps data using R. In *18th Euro Working Group on Transportation, EWGT*, 10, 276–285. The Netherlands: Transportation Research Procedia.
12. Patil, Swapnil, Milo Polte, Kai Ren, Wittawat Tantisiriroj, Lin Xiao, J. López, Garth Gibson, Adam Fuchs, and Billie Rinaldi. 2011. Ycsb++: benchmarking and performance debugging advanced features in scalable table stores. In *Proceedings of the 2nd ACM Symposium on Cloud Computing*, 9. ACM.
13. Sen, R., A. Farris, and P. Guerra. 2013. Benchmarking apache accumulo bigdata distributed table store using its continuous test suite. In *2013 IEEE International Congress on Big Data*, 334–341.
14. Sivaraman, S., and M.M. Trivedi. 2013. Looking at vehicles on the road: A survey of vision-based vehicle detection, tracking, and behavior analysis. *IEEE Transactions on Intelligent Transportation Systems* 14 (4): 1773–1795.

Exploiting Recurrent Neural Networks for Gate Traffic Prediction

Fabio Fumarola and Pasqua Fabiana Lanotte

Abstract Traffic information plays a significant role in everyday activities. It can be used in the context of *smart traffic management* for detecting traffic congestions, incidents and other critical events. While there are numerous ways for drivers to find out where there is a traffic jam at a given moment, the estimation of the future traffic is not used for proactive activities such as ensuring a smoother traffic flow and to be prepared for critical situations. Therefore traffic prediction is focal both for public administrations and for the Police Force in order to do resource management, network security and to improve transportation infrastructure planning. A number of models and algorithms were applied to traffic prediction and achieved good results. Many of them require the length of past data to be predefined and static, do not take into account dynamic time lags and temporal autocorrelation. To address these issues in this paper we explore the usage of Artificial Neural Networks. We show how Long Short-Term Memory (LSTM), a particular type of Recurrent Neural Network (RNN), can overcome the above described issues. We compare LSTM with a standard Feed Forward Neural Network (FDNN), showing that the proposed model achieves higher accuracy and generalises well.

1 Introduction

With the increasing amount of traffic information collected through automatic number plate reading systems (NPRS), it is highly desirable for public administrations and the *Police Force* to have surveillance tools to estimate traffic parameters such as the number of car passing through a plate or their average speed to detect traffic congestions, incidents and other critical events. For example, commercial traffic data providers, such as Bing maps Microsoft Research [13], rely on traffic flow data, and

F. Fumarola (✉) · P. F. Lanotte
University of Bari Aldo Moro, Bari, Italy
e-mail: fabio.fumarola@uniba.it

P. F. Lanotte
e-mail: pasqua.lanotte@uniba.it

© Springer International Publishing AG, part of Springer Nature 2018   145
F. Leuzzi and S. Ferilli (eds.), *Traffic Mining Applied to Police
Activities*, Advances in Intelligent Systems and Computing 728,
https://doi.org/10.1007/978-3-319-75608-0_11

machine learning to predict speeds given road segment. Real-time (15–40 min) forecasting gives travellers the ability to choose better routes and authorities the ability to manage the transportation system. While, for the Police Force it can be useful to plan surveillance activities. The predictability of network traffic parameters is mainly determined by their statistical characteristics and by the fact that they present a strong correlation between chronologically ordered values. Network traffic is identified by: self-similarity, long-range dependence and a highly nonlinear nature (insufficiently modelled by Poisson and Gaussian models).

Several methods have been proposed in the literature for the task of *traffic forecasting*, which can be categorised into: linear prediction and nonlinear prediction. The most widely used traditional linear prediction methods are: a) the ARMA/ARIMA model [4, 5, 8]. The most common nonlinear forecasting methods involve Artificial Neural Networks (ANNs) [1, 4, 6]. The experimental results from [3] show that nonlinear traffic prediction based on ANNs outperforms linear forecasting models (e.g. ARMA, ARAR, HW) which cannot meet the accuracy requirements.

Also if it has been proved that ANNs achieve the best results, deciding the best ANN architecture for the task is still a daunting task. The best architecture can be a compromise between the complexity of the solution, characteristics of the data and the desired prediction accuracy. Most of the times a simple Feed Forward Deep Neural Network (FFDNN) can achieve good results. But, we are going to show that by taking into account the implicit temporal characteristics of the analysed problem we can achieve better results.

Unlike feed forward deep neural networks (FFNNs), Recurrent Neural Networks (RNNs) have cyclic connections over time. The activations from each time step are stored in the internal state of the network to provide a temporal memory. This capability makes RNNs better suited for sequence modelling tasks such as time series prediction and sequence labeling tasks. However, RNNs suffer the well know problem of the *vanishing gradient* [7], that happens when multiplying several small values from the temporal memory. This makes RNNs not able to learn long temporal dependencies. On the contrary, Long Short-Term Memory (LSTM) [7] addresses the vanishing gradient problem of conventional RNNs by using an internal memory, a *carry* and a *forget* gate to decide when keep/forget the information stored in its internal memory. RNNs and LSTMs have been successfully used for handwriting recognition [10], language modelling [12], speech to voice [2] and other classification and prediction tasks.

In this paper we present a deep neural network (DNN) architecture based on LSTM to forecast hour by hour the number of vehicles passing through a gate. We build a general model that is capable of predicting the next 24 hours of traffic basing on the past observed data. Moreover, the model is capable of abstracting on the gate number and on the temporal component. Thus, it can do predictions for variable time ranges. We show the advantages of the proposed architecture in modelling temporal correlations with respect to a FFDNN in term of mean square errors. In remainder of the paper we discuss: RNN and LSTM, the proposed architecture for the task, the experimental results and we draw conclusions and discuss on future works.

2   RNN and LSTM

Recurrent Neural Network. Feed Forward Neural Networks treat each input as independent with respect the previous ones. They start for scratch for every input not taking into account possible correlation stored in consecutive observations of the input. On the contrary, RNNs address this issue, by using a loop (see Fig. 1) in them that allows information to flow.

Given an input-length l, the loop in the RNN unrolls the network l-*times* and creates multiple copies of the same network, each passing its computed state value to a successor (see Fig. 2).

The simple architecture of RNN has an input layer **x**, hidden layer **h** and output layer **y**. At each time step t, the values of each layer are computed as follows:

$$h_t = f(U_{x_t} + W_{h_{t-1}}) \tag{1}$$

$$y_t = g(V h_t) \tag{2}$$

where **U**, **W** and **V** are the connection weight matrices in RNN, and $f(z)$ and $g(z)$ are *sigmoid* and *softmax* activation functions.

Long Short-Term Memory. Long short-term memory (LSTM) [7] is a variant of RNN which is designed to deal with the gradient vanishing and exploding problem when learning with long-range sequences. LSTM networks are the same as RNN, except that the hidden layer updates are replaced by memory cells. Basically, a memory cell unit is composed of three multiplicative gates that control the proportions of information to forget and to pass on to the next time step. As a result, it is better for exploiting long-range dependency data. The memory cell is computed as follows:

$$i_t = \sigma(W_i h_{t_1} + U_i x_t + b_i) \tag{3}$$

$$f_t = \sigma(W_f h_{t_1} + U_f x_t + b_f) \tag{4}$$

$$c_t = f_t \odot c_{t-1} + i_t \odot tanh(W_c h_{t-1} + U_c x_t + b_c) \tag{5}$$

$$o_t = \sigma(W_o h_{t-1} + U_o x_t + b_o) \tag{6}$$

Fig. 1 Loops in Recurrent Neural Networks

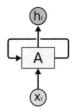

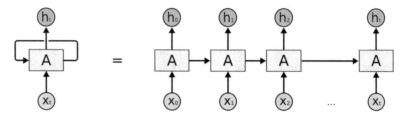

Fig. 2 Unrolled Recurrent Neural Networks

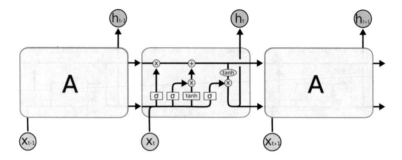

Fig. 3 The repeating module in an LSTM contains four interacting layers

$$h_t = o_t \odot tanh(c_t) \tag{7}$$

where σ is the element-wise sigmoid function and $\odot$ is the element-wise product, **i, f, o** and **c** are the input gate, forget gate, output gate and cell vector respectively. W_i, W_f, W_c, W_o are connection weight matrices between input **x** and gates, and U_i, U_f, U_c, U_o are connection weight matrices between gates and hidden state h. While, b_i, b_f, b_c, b_o are the bias vectors. Figure 3 shows how the information carried in the memory cell are modified and passed out the next LSTM cell.

3   Gate Traffic Prediction

In this section we describe the use of a LSTM to extract the temporal feature of the network traffic and predict hour by hour the number of vehicles passing through gates. This architecture can deeply model mutual dependence among gate measure entries in various time-slots.

3.1   Problem Statement

Let G be the list of the gates in a roads network, and T the time dimension, we can collect data from the gates at each timestamp t_i. Each measure contains information related to the gate number, the number of cars pass through, and the actual timestamp.

We can represent the data as a matrix of shape $L \times M$, where L is the number of measures and M is the number of features recorded. Given a list of consecutive measures for a gate we can model them as a time-series. The main challenge here is how to sample properly the batches of data to feed the LSTM for the training phase.

3.2 Modelling the Input for the LSTM

As input for the LSTM we draw a sliding window of measures extracted from the same gate. Each batch has:

- a **batch size**: the number of observations to generate
- a **sample size**: the number of consecutive observations to use as features
- a **predict step**: the position of the observation to use as target value for the prediction task.

Figure 4 shows an example of measure with: batch size equal to 3, sample size equal to 10 and predict step with value 1.

During the training we feed the LSTM with batches of *sample size* larger than the one showed in the example above. In particular, we are interested to predict the traffic state for the next 24 hours, thus we set this as fixed value for the evaluation.

3.3 Performance Metric

To quantitatively assess the overall performance of our LSTM model, Mean Absolute Error (MAE) is used to estimate the prediction accuracy. MAE is a scale dependent metric which quantifies the difference between the forecasted values and the actual values of the quantity being predicted by computing the average sum of absolute errors:

$$MAE = \frac{1}{N} \sum_{i=1}^{N} (|y_i - \hat{y}|) \tag{8}$$

where y_i is the observed value, $\hat{y}_i$ is the predicted value and N represents the total number of predictions.

Fig. 4 Example of sliding window batch

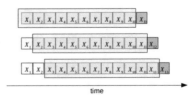

4   Experimental Results

In order to evaluate the proposed approach, we performed an experimental evaluation on **Trap-2017**[1] dataset. This dataset contains transits recorded using several gates. It stores measures from the transits of a limited area of Italy, in which gates are homogeneously distributed. The plates are coherently anonymised (a plate is always referred using the same ID). Data records contains: *plate id, gate number, street line number, timestamp, nation of the car*. All The data are split into a file per day and the whole dataset contains measured sampled for the 2016, which are in total 155.586.309 observations.

Since our goal to create a model to predict the number of cars flowing through a specific gate hour by hour, we preprocessed the data to obtain the following features: *gate number, hour, day, month, week-day, and the count of transit in 60 min range*. After this data transformation step the size of the dataset was reduced to 222.115 observations. This is mostly because we aggregate the measures to 60 min, if data was aggregated each 15 min we would have a size four times larger.

We compare results for the proposed LSTM with respect a Feed Forward Deep Neural Network architecture (FFDNN). *Mean Square Error* (see Eq. 9) is used as performance measure. To implement both the models we used Keras[2] Deep Learning Library. The first architecture is composed by 4 LSTM layers with 64, 128, 64 and 1 neurons per layer, while the second uses 4 Dense Layers with 64, 128, 64 and 1 neurons. As activation functions we used Rectified Linear Unit (ReLU) [9], while as optimizer RMSprop [11]. The dataset is split in 70%, 10%, 20% for training, validation and test.

Mean Square Error (MSE) (Eq. 9) is used as loss function for the optimizer. MSE is a scale dependent metric which quantifies the difference between the forecasted values and the actual values of the quantity being predicted by computing the average sum of squared errors:

Fig. 5 Training and validation mean absolute error for the FFDNN

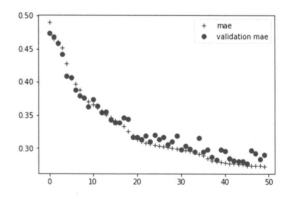

[1] https://trap2017.poliziadistato.it/.
[2] http://keras.io/.

$$MSE = \frac{1}{N} \sum_{i=1}^{N} (y_i - \hat{y})^2 \qquad (9)$$

where y_i is the observed value, $\hat{y}_i$ is the predicted value and N represents the total number of predictions.

The FFDNN and the LSTM models were trained for 50 epochs. Figures 5 and 6 present training and validation MAE for the FFDNN. From the results is interesting to see that the architecture converge nicely without overfitting. After the training the minimum MAE for the validation set is 0.29, while for the test set is 0.287.

On the contrary, Figs. 7 and 8 present the results for the LSTM. Here we can note that the convergence rate is faster than the one observed for the FFDNN network. Finally, it achieves 0.121 and 0.134 of MAE score for the validation and test set respectively, which is half the of the value obtained for the FFDNN.

In Figs. 9 and 10 we can see an example of prediction done by the two models. Here we can note that the prediction made by the LSTM is smoother than the one made by the FFDNN.

Fig. 6 Training and validation loss for the FFDNN

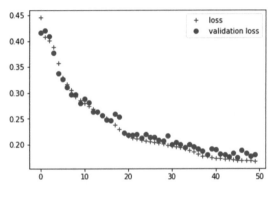

Fig. 7 Training and validation mean absolute error for the LSTM

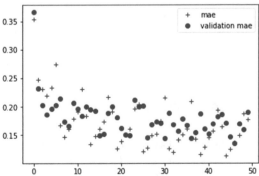

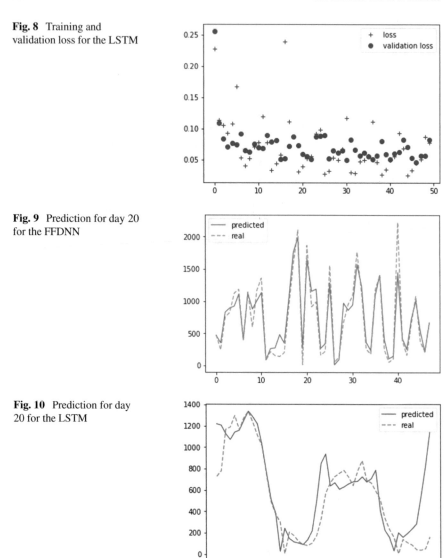

Fig. 8 Training and validation loss for the LSTM

Fig. 9 Prediction for day 20 for the FFDNN

Fig. 10 Prediction for day 20 for the LSTM

5 Conclusions and Future Works

In this paper, we have presented a new neural network architecture based on LSTM for gate traffic prediction. Experiments results show that the proposed architecture performs well on the Trap-2017 dataset in terms of mean absolute error, and that it obtains better result with respect to a classical FFDNN. As future work we will run our algorithm on different datasets and do an extensive search for hyper-parameters tuning.

References

1. Abdennour, A. 2006. Evaluation of neural network architectures for mpeg-4 video traffic prediction. *IEEE Transactions on Broadcasting* 52 (2): 184–192.
2. Arik, Sercan Ömer, Mike Chrzanowski, Adam Coates, Greg Diamos, Andrew Gibiansky, Yongguo Kang, Xian Li, John Miller, Jonathan Raiman, Shubho Sengupta, and Mohammad Shoeybi. Deep voice: Real-time neural text-to-speech. *CoRR*, abs/1702.07825, 2017.
3. Barabas, M., G. Boanea, A. B. Rus, V. Dobrota, and J. Domingo-Pascual. 2011. Evaluation of network traffic prediction based on neural networks with multi-task learning and multiresolution decomposition. In *2011 IEEE 7th International Conference on Intelligent Computer Communication and Processing*, 95–102.
4. Cortez, Paulo, Miguel Rio, Pedro Sousa, and Miguel Rocha. 2007. *Topology Aware Internet Traffic Forecasting Using Neural Networks*, 445–454. Berlin, Heidelberg: Springer.
5. Dai, J., and J. Li. 2009. Vbr mpeg video traffic dynamic prediction based on the modeling and forecast of time series. In *2009 Fifth International Joint Conference on INC, IMS and IDC*, 1752–1757.
6. Dharmadhikari, V.B., and J.D. Gavade. 2010. An nn approach for mpeg video traffic prediction. In *2010 2nd International Conference on Software Technology and Engineering*, vol. 1, pp. V1–57–V1–61.
7. Hochreiter, Sepp, and Jürgen Schmidhuber. 1997. Long short-term memory. *Neural Computation* 9 (8): 1735–1780.
8. Joshi, Manish, and Theyazn Hassn Hadi. 2015. A review of network traffic analysis and prediction techniques. *CoRR*, abs/1507.05722.
9. Nair, Vinod, and Geoffrey E. Hinton. 2010. Rectified linear units improve restricted boltzmann machines. In *Proceedings of the 27th International Conference on International Conference on Machine Learning*, ICML'10, 807–814, USA, Omnipress.
10. Stuner, Bruno, Clément Chatelain, and Thierry Paquet. 2016. Cohort of LSTM and lexicon verification for handwriting recognition with gigantic lexicon. *CoRR*, abs/1612.07528.
11. Tieleman, T., and G. Hinton. 2012. Lecture 6.5—RmsProp: Divide the gradient by a running average of its recent magnitude. COURSERA: Neural Networks for Machine Learning.
12. Verwimp, Lyan, Joris Pelemans, Hugo Van hamme, and Patrick Wambacq. 2017. Character-word LSTM language models. *CoRR*, abs/1704.02813.
13. Zhang, Junbo, Yu Zheng, and Dekang Qi. 2016. Deep spatio-temporal residual networks for citywide crowd flows prediction. AAAI 2017.

Author Index

© Springer International Publishing AG, part of Springer Nature 2018
F. Leuzzi and S. Ferilli (eds.), *Traffic Mining Applied to Police
Activities*, Advances in Intelligent Systems and Computing 728,
https://doi.org/10.1007/978-3-319-75608-0

155